Lecture Notes in
Computer Science

T0216479

Lecture Notes in Computer Science

Lecture Notes in Computer Science

Edited by G. Goos and J. Hartmanis

474

D. Karagiannis (Ed.)

Information Systems and Artificial Intelligence: Integration Aspects

First Workshop
Ulm, FRG, March 19–21, 1990
Proceedings

Springer-Verlag
Berlin Heidelberg New York London
Paris Tokyo Hong Kong Barcelona

Volume Editor

Dimitris Karagiannis
Artificial Intelligence Laboratory
Forschungsinstitut für anwendungsorientierte Wissensverarbeitung (FAW)
Universität Ulm
Postfach 2060, W-7900 Ulm, FRG

CR Subject Classification (1987): H.2, I.2.3–5

ISBN 3-540-53557-8 Springer-Verlag Berlin Heidelberg New York
ISBN 0-387-53557-8 Springer-Verlag New York Berlin Heidelberg

© Springer-Verlag Berlin Heidelberg 1991
Printed in Germany

Printing and binding: Druckhaus Beltz, Hemsbach/Bergstr.
2145/3140-543210 – Printed on acid-free paper

Preface

Es ist nicht genug zu wissen,
man muß auch anwenden;
es ist nicht genug zu wollen,
man muß auch tun.

Goethe

The First Workshop on "Information Systems and Artificial Intelligence" took place at the FAW Ulm, March 19–21, 1990. Its focus was on "Information- and Knowledge-Representation Systems". The special interest groups 1.1.4 (Knowledge Representation) and 2.5.2 (Methods for Developing Information Systems and Their Application) of the GI e.V. (German Computer Science Society) were responsible for the organization.

Knowledge-based systems have been successfully developed in practice for a number of years. However, they are often "only" stand-alone systems; integrating them into existing information environments, e.g. making available real production data to an expert system, often either fails or is only solved in a dissatisfying way. Possible reasons for this might be on one hand the lack of know-how about the different features of various experimental AI techniques, and on the other the lack of more classical information and database system technology. The possibility for change led the special interest groups "Knowledge Representation" and "Methods for the Development of Information Systems and their Application" to organize a joint workshop entitled *"Information Systems and Artificial Intelligence: Integration Aspects"*.
The papers in this volume illustrate how approaches integrating Artificial Intelligence and Database technology may be developed.

Database and Artificial Intelligence Techniques: A New Approach Towards Information Systems?

Information systems are not only characterized by their conceptual design, but also by their application domain. New technologies and systems resulting from database and expert systems have influenced methods of analysis, design and specification, leading to realization approaches which play an important role in the development of information systems.
When developing and realizing expert and/or database systems, differences can be found at both the system-technical level and at the conceptual level.
For the integration of expert and database systems into information systems, basically two strategies are suggested in the literature:

a) coupling of both systems, and
b) extension of existing technologies.

Coupling Approaches

The objective of coupling approaches is to enable autonomous systems — in this case restricted to database and expert systems — to communicate in order to increase their efficiency and application adequacy with regard to joint problem-solving. The requirements for such a coupling are usually realized via "interface" concepts. Different coupling approaches can be found in the literature; we distinguish approaches with distinct database and distinct expert system aspects respectively.

An alternative to coupling approaches is the extension of existing systems with additional abilities. Such extensions may lead to better performance of the resulting systems and are referred to here as "additive approaches".

Additive Approaches

Usually, what is described is the development of a new DB model which is either able to manage complex structures or has been extended by data models and processing facilities based on a resolution calculus; it might also include heuristic processing mechanisms.

In the area of Expert Database Systems — which can be regarded as the bottom-up approach for realizing knowledge-based systems — it is rather AI methods that are applied. Extension approaches can be classified into

a) approaches in which the existing (logic) programming environment is extended with database facilities, and

b) approaches in which a database system is extended with deductive abilities.

Since the different extension approaches come primarily from database technology, there exists a conceptual distance from AI technologies. This becomes clear when dealing with knowledge representation. A great number of articles on this subject can be found in current publications.

"Non-classical" Information Systems

"Non-classical" information systems are meant to build up a new class of systems, using an interdisciplinary approach allowing an assignment within the framework of artificial intelligence and the further development of data models. These are tailored with regard to the applications which are to be realized. At the moment there exists neither a procedure for constructing these systems nor a hypothetical system architecture. This type of information system should combine expert and database system functionality.

Due to the requirements that such systems pose, they can only be realized in an interdisciplinary fashion. Although this workshop focusses on Database- and Artificial Intelligence techniques, aspects of software engineering and insights from programming language disciplines have also to be taken into consideration. It will have to be analysed whether "non-classical" information systems within this context provide research potential for the next years. Different approaches in this direction are also described in the current literature.

Scope of the Papers

The basic consideration was to have two representatives — one from the Artificial Intelligence and one from the Database Systems field — talk on the topic determined by each respective section. The different views were meant to generate the basis of potential ideas and to support the discussion.

With regard to the contents, the workshop was divided into 6 sections with 18 talks. A panel session on experiences from practice resulted in additional ideas to be discussed.

The workshop started with an overview talk on the topic **Databases and Artificial Intelligence** given by the invited speaker John Mylopoulos of the University of Toronto. The specific topics covered in the workshop were:

- Logic Programming and Deductive Databases
- Non-Monotonic Logic
- Conceptual Model and Knowledge Representation
- Database Support for Knowledge-Based Systems
- Database Systems and Expert Systems: A New Approach for Information Systems?
- Requirements Analysis vs. Knowledge Acquisition

Logic Programming and Deductive Databases: Stefan Böttcher of IBM Deutschland GmbH in Stuttgart described the integration of a deductive database with logic programming being pursued in the EUREKA project PROTOS. The logic programming language PROTOS-L (developed for that purpose) was explained as well.

Rainer Manthey of the ECRC in Munich talked about two basic problems; for their solution it was necessary to rely upon terms and methods of database technology:

a) the efficient checking of integrity conditions, and
b) the efficient answering of queries on recursive rules.

Non-Monotonic Logic: Gerhard Brewka of the GMD in St. Augustin described the approaches of non-monotonic reasoning and explained the advantages and disadvantages.

François Bry of the ECRC in Munich explained the necessity to continue working with other kinds of logic, e. g., constructive logic. He reported about the insights gained within the framework of the ESPRIT projects, such as formalizing in logic programming, handling of negation and generating implementation approaches for models of non-monotonic logic and default reasoning in logic programming.

Conceptual Model and Knowledge Representation: Christof Peltason of the TU Berlin explained existing possibilities and experiences gained in connection with the realization of the knowledge representation system BACK (Berlin Advanced Computational Knowledge Representation System) within the framework of an ESPRIT project.

Bernhard Nebel of the DFKI in Saarbrücken extended the area of knowledge representation and talked about problems and approaches to solutions. Finally, he stated that knowledge representation formalisms are only useful when the processing of the represented knowledge may be efficiently realized — actually the problem is not completely solved.

Gunther Saake of the University of Braunschweig presented a layer model for the conceptual design. This shall help the software engineer to correctly interpret the user's requests. He sees a realization approach for his concept in the area of object-oriented databases.

Database Support for Knowledge-Based Systems: Günther Görz of the University of Hamburg emphasized the importance of database support for knowledge representation.

Bernd Walter of the University of Trier mentioned some basic considerations regarding database support for knowledge-based systems; in addition he presented the approaches pursued in the LILOG database.

Database Systems and Expert Systems: A New Approach for Information Systems? Dieter Hovekamp of VW-Gedas Berlin described how a KBMS was implemented in a PC environment; this was effected within the framework of the BMFT project WEREX.

Nelson Mattos of the University of Kaiserslautern presented an approach for realizing a prototype database-based knowledge management system. The functionality description of the KRISYS system and the differences from non-standard database approaches were presented as well.

Requirements Analysis vs. Knowledge Acquisition: Angela Voß of the GMD St. Augustin gave an overview of model-based knowledge acquisition. Different advantages and disadvantages as well as comparisons with conventional software development were discussed.

Helmut Thoma of Ciba Geigy AG in Basel, Switzerland reported on approaches to the information requirement analysis problem. His talk was influenced by practice and argued for a gradual build-up of knowledge and requirements in concrete form during system development.

Panel Session

In the discussion "Experiences from Practice", many participants of the workshop discussed the difficulties with which practice is still nowadays struggling in integrating information systems and expert systems. The session was chaired by Helmut Thoma, Ciba Geigy AG Basel, and he and Hans Peter Hoidn, Institut für Automation AG Zürich, Nelson Mattos, University of Kaiserslautern and Wolfgang Sager, Collogia Köln determined the main focal points and answered the questions arising.

H.-P. Hoidn reported on the practical aspects of implementing a knowledge base/database coupling in a major bank. They implemented a system for planning the batch control in the computer centre of the bank with KEE and Common Windows, with ORACLE and KEE-Connection for coupling object- and database systems. Experiences with performance evaluation and analyses of coupled (relational) database and expert systems, as they can hardly be effected in practice, were at the centre in attention of N. Mattos' explanations. In enterprises, knowledge-based systems have to be integrated on the technical, the management and the user level as realizations of innovative techniques in existing environments. W. Sager discussed those points of view in connection with his experience with AI techniques in the conventional environments within companies.

Nowadays the state of the art unfortunately comprises loose couplings of artificial intelligence systems and conventional systems, problems with integrated development tools, with too few project leaders or employees comprehending both techniques. In addition, the effort needed to actualize knowledge in the working phase is often underestimated.

The design of object classes and relations should start with a common model as the basis, e. g., a common entity relationship model. Special attention has to be given to the integrity of data, to the object system, and to the conception of transactions.

Expert Systems manipulate the data in a way that is unfavourable for database systems. Therefore, on coupling, the locality of processing should be better used, storage structures of the Database System should be better adapted to the working method of the Expert Systems, and Expert Systems calls should be adapted to the quantity-oriented interface of the Database System. Furthermore, an optimal Expert Systems/Database System coupling requires a distribution of tasks and design of interfaces that are well suited to each other; this could be proven in experiments.

Conclusions

As an exchange of experiences, the workshop was an important step forward to a mutual understanding between persons who are occupied in practice with information systems and database systems, and their colleagues working in the area of Artificial Intelligence, especially knowledge representation. The terminological as well as the conceptual differences between both areas could be seen in several discussions.

The consensus that emerged was that both groups should make steps to fulfill the existing requirements on new software systems by cooperation. Future developments will determine whether this leads to interdisciplinary work and real integration approaches.

Acknowledgement

I would like to thank those of my colleagues who helped to set up this workshop, and in particular the members of the organizing committee, G. Lausen, University of Mannheim, K. von Luck, C.-R. Rollinger, IBM Deutschland GmbH, R. Studer, University of Karlsruhe, and H. Thoma, Ciba-Geigy AG. Thanks go also to the people who helped in the local organisation at the FAW Ulm.

Whether this line of research bears fruit, only time can tell. But, whatever the outcome, I owe an inestimable debt to all participants for their good will and manifold assistance.

Ulm, October 1990   Dimitris Karagiannis

Contents

A semantics for the integration of database modifications and transaction brackets into a logic programming language

Stefan Böttcher

IBM Deutschland GmbH
Scientific Center
Institute for Knowledge Based Systems
P.O.Box 80 08 80
D – 7000 Stuttgart 80
West Germany [1]

Abstract

The integration of logic programming and databases has up to now focussed on read access of logic programs to external databases storing permanent data. However, the integration of write operations modifying existing databases would allow to use logic programming languages in a much larger field of applications. Therefore, it is important that the logic programming language not only embeds modification operations on existing databases, but also embeds transactions in order to preserve the correctness of the modified database.

This paper describes an integration of database modifications and transactions into a logic programming language evaluated by a depth first left to right strategy with backtracking. We propose a semantics of insert and delete operations and outline why these operations are defined different from assert and retract in Prolog. Furthermore, we propose a semantics for transaction brackets and describe how these operations differ from begin_transaction and end_transaction statements in database programming languages.

---

[1]The research reported here has been carried out within the international EUREKA project PROTOS (EU56): Prolog Tools for Building Expert Systems.

1 Introduction

During the last years, the integration of logic programming and databases has become an increasingly important research area, because the integration enables logic programs to use the knowledge stored in existing databases. However, in order to fully support high-level database access from logic programs, the integration of database modifications has been considered to be an important extension of logic programming (e.g. [Fagin *et al.*, 1986], [Manchanda and Warren, 1988], [Wilkins, 1986]). In contrast to these approaches we argue that correctness of database updates additionally requires to integrate a transaction concept and that this may lead to a different judgement about update semantics.

While the integration of database modifications and transactions into procedural programming languages as e.g. Pascal or Modula-2 is well understood [Schmidt, 1977], [Schmidt *et al.*, 1988], the integration of these concepts into logic programming languages leads to the following problem: How shall we integrate database modifications and a transaction concept into the backtracking evaluation strategy used for logic programming languages ?

The problem can be devided into two subproblems: first, to find a clean integration of database modifications avoiding "dirty side-effects", and second, to integrate a transaction concept into backtracking such that atomicity and persistence of transaction executions are not violated, when several transactions are executed by one single logic program.

In this paper we describe a proposal to solve this problem based on the logic programming language PROTOS-L, which is currently developed at IBM Scientific Center in Stuttgart as part of the EUREKA project PROTOS (EU 56). PROTOS-L [Beierle, 1989], [Beierle and Böttcher, 1989] is a logic programming language, providing access to external databases, a polymorphic order-sorted type concept, and a module concept similar to that of Modula-2 [Wirth, 1983]. A compiler and an abstract machine[2] for PROTOS-L have been implemented on the IBM-RT 6150 workstation [Semle, 1989].

PROTOS-L is currently used to reimplement a part of an algorithm which is used in a prototype of a chemical production planning system for Sandoz AG. This production planning system uses large sets of heuristic rules coded in logic and needs access to planning data which is stored in a relational database system. Therefore it requires to embed database access into a logic programming language.

The next section summarizes the requirements, while the third section describes a proposal for the integration of database modification operations and a transaction concept into the logic programming language PROTOS-L.

---

[2]The abstract machine for PROTOS-L is an extension of the Warren Abstract Machine [Warren, 1983].

2 Requirements for the integration of transactions and database modifications into logic database programming languages

In this section we describe the requirements for the integration of transactions and database modifications into a logic programming language. The description focusses on how a given logic programming language can be extended to a logic database programming language. The requirements for the integration of transactions and backtracking will lead to corresponding requirements for the integration of database modifications and backtracking.

2.1 Requirements for the integration of transactions and backtracking

The integration of transactions into a logic database programming language has some requirements to both the embedding language and program execution. The embedding logic programming language shall have the following properties in order to integrate transactions:

1. The code of several transactions can be combined in one single logic program.

2. Transaction steps are programmed in the embedding logic programming language.

3. Integrity checking can be programmed in the embedding logic programming language.

The basic requirements to integrate transaction executions and backtracking are the following:

4. Transaction executions have to be atomic, i.e. atomicity of transaction executions has to be embedded in backtracking.

5. Database modifications of committed transaction executions have to be persistent, i.e. a concept of transaction persistence has to be integrated with backtracking.

Reqiurement 4 implies that all modification operations executed in a transaction have to be undone in case of a transaction abort. In order to meet this requirement, we will propose a semantics of write operations which is free of side effects in this case (c.f. sections 3.1 and 3.3).

Further, both atomicity and persistence of transaction executions require that backtracking is prevented from jumping inside an already committed transaction (c.f. section 3.2).

2.2 The integration of write operations and backtracking

The requirements for the integration of write operations and backtracking can be derived from the requirements 4 and 5 for the integration of transactions and backtracking:

6. Write operations are undone if the transaction they occur in is aborted, i.e. we want to avoid that write operations have side-effects that survive the transaction abort.

7. Write operations are made permanent if the transaction they occur in is committed.

Finally, we require what follows in order to keep programming in the language simple:

8. Write operations do not influence the flow of program execution, i.e. write operations are always successful (c.f. section 3.1).

3 A solution based on PROTOS-L

PROTOS-L [Beierle, 1989], [Beierle and Böttcher, 1989] is a typed logic programming language which is developed and implemented at IBM Stuttgart. PROTOS-L embeds read access to external databases, a module concept similar to that of Modula-2, and a polymorphic order-sorted type system supporting fast program execution. A detailed description of PROTOS-L as a database query language is given in [Böttcher, 1990a], whereas the PROTOS-L run time system is described e.g. in [Beierle, 1989], [Böttcher and Beierle, 1989] and [Böttcher, 1990b].

In this paper we concentrate on a proposal for the integration of database modification operations and a transaction concept into the programming language PROTOS-L. Subsection 3.1 and 3.3 describe how database modifications can be embedded, while subsections 3.2 and 3.4 outline how the suggested extension of PROTOS-L integrates transactions, in order to meet the requirements summarized in the last section.

3.1 A proposal for database modifications in the programming language PROTOS-L

Under the aspect of database access, PROTOS-L is basically intended to be a database query language, not to be a database modification langauge. Nevertheless, insert (:+) and delete (:\) operations can be embedded in the programming language PROTOS-L as follows. Insert and delete operations are goals, i.e. they can occur in any right hand side of a rule.

In order to describe the semantics of insert and delete operations we outline in this section,

- what are the preconditions for the operations,

- whether these operations sometimes fail or they have always success, and

- how the database state is changed when program execution passes these operations from left to right.

In section 3.3 we complete this description by explaining how the database state is changed when backtracking passes these operations from right to left.

We use the following example instead of giving a full set of definitions.

Let R be a database relation and $A_1, \ldots, A_n$ be arguments, i.e. A_i is either a variable or a constant. The precondition for the execution of the operations

$$R :+ (\ A_1, \ldots, A_n\ ) \quad \text{and} \quad R :\backslash (\ A_1, \ldots, A_n\ )$$

is that each argument which is a variable is bound to a constant at the calling time of the operation. Hence, at calling time all arguments are bound to constants, say $c_1, \ldots, c_n$.

We further have to decide whether a delete operation (and an insert operation respectively) is always successful, or it is only successful if the tuple to be deleted (inserted) exists (does not exist) in the database relation. In order to meet requirement 8, insert and delete statements are always successful in the suggested extension of PROTOS-L, independent of the tuples currently stored in the database relation R.

The database state after the execution of the write operation (by passing it from left to right) is the same as the database state before the execution of the write operation with the following exception: If R_{pre} is the value of the relation R before the execution of a write operation, then

$$R_{pre} \cup \{(c_1, \ldots, c_n)\}$$

is the value of the relation R after the execution of the insert operation and

$$R_{pre} \setminus \{(c_1, \ldots, c_n)\}$$

is the value of the relation R after the execution of the delete operation.

The description how the database state is changed when backtracking passes database modifications will be delayed, until we have described the PROTOS-L transaction concept.

Finally, we give an example for a delete and an insert operation from a program part of our chemical production planning application. An execution of the following database rule changes the location, where a (chemical) product P is planned to be stored within the time interval [T1 , T2], from location1 to location2.

```
change_location ( P , T1 , T2 , 'location1' , 'location2' ) :-
    product_is_at :\ ( P , T1 , T2 , 'location1' ) ,
    product_is_at :+ ( P , T1 , T2 , 'location2' ) .
```

3.2   A proposal for transactions

In order to describe the scope of transactions, we suggest that the programming language PROTOS-L offers transaction brackets ({ and }) as language constructs, i.e. the action sequence of the transaction is enclosed in these transaction brackets. Note however, that the meaning of the transaction brackets ({ and }) differs from the meaning of the begin_transaction and end_transaction statements within procedural database programming languages, e.g. DBPL [Böttcher, 1989]. Further, the transaction brackets ({ and }) influence the flow of program execution and backtracking in order to meet requirements 4 and 5. We give the following example from our chemical production planning application[3] in order to discuss both aspects of the suggested transaction concept.

The transaction change_if_legal changes the planned storage location of a chemical product for the time intervall [T1,T2] from location L1 to location L2, only if the succeeding integrity_checks are successful.

```
change_if_legal(P,T1,T2,L1,L2) :-
    { , change_location(P,T1,T2,L1,L2) ,
      integrity_checks(P,T1,T2,L1,L2) , } .
```

After changing the location the integrity checks are performed. If they are successful, then the transaction is committed as soon as the } transaction bracket is passed from left to right. [4] On the other hand, if the integrity checks fail, then the transaction execution can not be completed successfully and the transaction is aborted when backtracking passes the left transaction bracket ({).

The integration of transaction brackets ({ and }) into backtracking described above can be summarized and generalized as follows:

- If program execution proceeds from left to right over a left transaction bracket ({), then the begin_of_transaction statement (tBegin for short) is performed.

- If program execution passes a left transaction bracket ({) from right to left (i.e. by backtracking), then the abort_of_transaction statement (tAbort for short) is performed.

---

[3]The requirements of this production planning application are summarized in [Böttcher, 1990c].
[4]At commit time the modifications of the transaction are made permanent to the database.

- If program execution proceeds from left to right over a right transaction bracket (}), then the commit_transaction statement (tCommit for short) is performed.

But, if program execution passes a right transaction bracket (}) from right to left (i.e. by backtracking), then the following problem arises: Backtracking from right to left over a right transaction bracket (}) has to be prevented from jumping to a choice point inside the transaction, because this transaction has already been committed.

The suggested solution to this problem is that, instead of jumping to a choice point inside a transaction backtracking has to jump to the last choice point allocated before the beginning of the transaction. Hence, the suggested solution meets requirement 4 for the integration of backtracking and transactions that transaction execution has to be atomic.

Like in other database programming languages, nested transaction calls are not allowed in this proposal. Since the modifications of transactions are made permanent to the database at commit time, and backtracking is prevented from jumping to a choice point inside a transaction, the proposed solution also meets requirement 5, that modifications of committed transactions are persistent.

3.3 The integration of database modifications and backtracking

In this setion we propose what has to be done, if backtracking goes from right to left over a database modification operation.

Requirement 6 states that all modifications executed in a transaction shall be undone, if the transaction is aborted, i.e. at the latest when backtracking reaches the left transaction bracket ({). The goal of this requirement is to avoid that write operations survive the transaction abort as side-effect of program execution.

The suggested solution to meet this requirement is as follows: Whenever backtracking passes a modification operation from right to left, then the modification operation is undone, i.e. backtracking reestablishes the database state given before this modification operation was executed. Hence, all database modifications are undone at transaction abort time, i.e. when backtracking reaches the left transaction bracket ({).

Note that insert and delete operations are different from assert and retract in Prolog [Cloksin and Mellish, 1981], because these modification operations are undone, as soon as backtracking passes these operations from right to left.

One the other hand, requirement 7 states that modifications of committed transactions have to be acchieved as follows: When program execution performs backtracking over a right transaction bracket (}), i.e. program execution goes to the last choice point activated before the transaction execution has begun, then the modifications made during transaction execution remain in the database.

To summarize: The only side-effect of write operations to the database during logic program execution is that side-effect which was required: modifications done in committed transactions are persistent.

3.4   A proposal for the implementation of transactions in the PROTOS-L compiler

The transaction concept can be implemented within the PROTOS-L compiler as follows. The compiler first translates PROTOS-L source code into some intermediate language, then it performs transformations and optimizations on this intermediate language representation of the program, and finally it translates the intermediate language representation into code for the PROTOS abstract machine.

At the intermediate language level, the code for transactions is transformed into logic programming code. The implementation of a transaction { ... } includes a cut in order to prevent backtracking from jumping from the outside of some transaction to a choice point allocated inside the transaction.

A transaction rule

```
transaction(P,T1,T2,L1,L2) :-
    { , transaction_body(P,T1,T2,L1,L2) , } .
```

can be implemented using the three built-ins tBegin, tCommit, and tAbort for the begin, the commit, and the abort of a transaction. These builts-ins are implemented by calls to the corresponding database system procedures of the underlying database system (which is in our case the SQL/RT database system). At the intermediate language level, the transaction rule is transformed into the following code:

```
transaction(P,T1,T2,L1,L2) :-
    transaction_implementation(P,T1,T2,L1,L2) .

transaction_implementation(P,T1,T2,L1,L2) :-
    tBegin , transaction_body(P,T1,T2,L1,L2) , tCommit , !  .

transaction_implementation(P,T1,T2,L1,L2) :- tAbort , fail .
```

Note that the cut behind the tCommit call prevents backtracking from jumping from outside the transaction to a choice point allocated by the transaction_body.

4 Summary and conclusion

The typed logic programming language PROTOS-L includes subtypes, polymorphism, a module concept and high-level read access to databases. We made a proposal how to integrate write operations and a transaction concept in order to extend this logic programming language to a database programming language. The transaction concept allows the application programmer to program transactions including the enforcement of integrity constraints within the logic programming language PROTOS-L. Several transactions can be combined in one single logic program. Further, the only side-effect of write operations during logic program execution is the side-effect which was required: modifications done in committed transactions are persistent. Hence, the suggested embedding of database modifications and transactions in the logic programming language PROTOS-L integrates backtracking and basic database programming concepts. The integration allows to implement integrated systems (using logic programming and embedding the knowledge of existing databases) in one single logic programming language. This opens up a large field of applications for logic programming languages.

References

[Beierle, 1989] C. Beierle. Types, modules and databases in the logic programming language PROTOS-L. In K. H. Bläsius, U. Hedtstück, and C.-R. Rollinger, editors, *Sorts and Types for Artificial Intelligence*, Springer-Verlag, Berlin, Heidelberg, New York, 1989. (to appear).

[Beierle and Böttcher, 1989] C. Beierle and S. Böttcher. PROTOS-L: Towards a knowledge base programming language. In *Proceedings 3. GI-Kongreß Wissensbasierte Systeme, Informatik Fachberichte*, Springer-Verlag, Berlin, Heidelberg, 1989.

[Böttcher, 1989] S. Böttcher. *Prädikative Selektion als Grundlage für Transaktionssynchronisation und Datenintegrität*. PhD thesis, FB Informatik, Univ. Frankfurt, 1989.

[Böttcher, 1990a] S. Böttcher. Development and programming of deductive databases with PROTOS-L. In L. Belady, editor, *Proc. 2nd International Conference on Software Engineering and Knowledge Engineering*, Skokie, Illinois, USA, 1990.

[Böttcher, 1990b] S. Böttcher. Integrating a deductive database system with a Warren Abstract Machine. In N. Cercone and F. Gardin, editors, *Proc. International Symposium Computational Intelligence 90*, Milan, Italy, 1990. (to appear).

[Böttcher, 1990c] S. Böttcher. A tool kit for knowledge based production planning systems. In M. Tjoa, editor, *Proc. International Conference on Data Base and Expert System Applications*, Springer-Verlag, Vienna, Austria, 1990. (to appear).

[Böttcher and Beierle, 1989] S. Böttcher and C. Beierle. Data base support for the PROTOS-L system. *Microprocessing and Microcomputing*, 27(1–5):25–30, August 1989.

[Cloksin and Mellish, 1981] W.F. Cloksin and C.S. Mellish. *Programming in Prolog*. Springer-Verlag, Berlin, Heidelberg, New York, 1981.

[Fagin *et al.*, 1986] R. Fagin, G.M. Kuper, J.D. Ullman, and M.Y. Vardi. Updating logical databases. In P. Kannellakis, editor, *Advances in Computing Research*, Jai Press, 1986.

[Manchanda and Warren, 1988] S. Manchanda and D.S. Warren. A logic-based language for database updates. In J. Minker, editor, *Foundations of Deductive Databases and Logic Programming*, Morgan Kaufmann, Los Altos, 1988.

[Schmidt, 1977] J.W. Schmidt. Some high level language constructs for data of type relation. *Transactions on Database Systems*, 2(3):247–261, 1977.

[Schmidt *et al.*, 1988] J.W. Schmidt, H. Eckhardt, and F. Matthes. *DBPL Report*. DBPL-Memo 111-88, Univ. Frankfurt, 1988.

[Semle, 1989] H. Semle. *Erweiterung einer abstrakten Maschine für ordnungssortiertes Prolog um die Behandlung polymorpher Sorten*. Diplomarbeit Nr. 583, Universität Stuttgart und IBM Deutschland GmbH, Stuttgart, April 1989.

[Warren, 1983] D. Warren. *An Abstract PROLOG Instruction Set*. Technical Report 309, SRI, 1983.

[Wilkins, 1986] M.W. Wilkins. A model-theoretic approach to updating logical databases. In *Proceedings of the 5th International Conference on Principles of Database Systems*, 1986.

[Wirth, 1983] N. Wirth. *Programming in Modula-2*. Springer, Berlin, Heidelberg, New York, 1983.

Handling Incomplete Knowledge
in Artificial Intelligence

Gerhard Brewka , GMD, Postfach 12 40

5205 Sankt Augustin, W. Germany

Abstract

In this paper we first discuss the important role of nonmonotonic reasoning for Artificial Intelligence. After presenting some simple forms of nonmonotonicity as they arise in various well-known AI systems we present in Section 2 some of the most important existing nonmonotonic logics: McCarthy's circumscription, Moore's autoepistemic logic, and Reiter's default logic. Section 3 examines an approach in which default reasoning is reduced to reasoning in the presence of inconsistent information. The approach is based on the notion of preferred maximal consistent subsets. It is shown that these preferred subsets can be defined in such a way that it is possibly to represent priorities between defaults adequately. Section 4 briefly discusses the problem of implementing nonmonotonic systems.

1. Introduction

Our empirical knowledge about the world is always incomplete and very often does not allow for the sound derivation of the conclusions necessary for decision-making, planning and acting. We have to act and decide in spite of this lack of knowledge, however. Very often we use rules with exceptions, which allow us to express what is typically the case, for filling the gaps in our knowledge. We use such default rules to derive plausible conclusions if there is no contradictory evidence. Every intelligent agent has to be able to handle these default rules in a reasonable way.

Another important ability of intelligent agents is handling inconsistent information. It happens very often that the information at hand is contradictory, for instance when it has been obtained from different, possibly unreliable sources. It is extremely important that intelligent agents do not become paralyzed by such situations.

For both problems, classical first order logic (FOL) has no solution to offer: in the case of inconsistent information, FOL derives the set of all formulas, which certainly is not what we want. Also, the handling of rules with exceptions cannot adequately be modelled in FOL. Of course, we can represent a default like "Birds typically fly" as follows:

(1) $\forall x.\text{BIRD}(x) \& \neg\text{EXCEPTION}(x) \rightarrow \text{FLIES}(x)$

(2) $\forall x.\text{EXCEPTION}(x) <\text{-}> \text{PENGUIN}(x) \lor \text{OSTRICH}(x) \lor \neg\text{HAS-WINGS}(x) ...$

This representation, however, requires a complete list of all possible exceptions, an impossible task in the light of the incompleteness of our knowledge. And even if such a complete description of exceptions were available, the representation would still be inadequate: in order to derive for a particular bird, say Tweety, that it flies, it is necessary to prove that no exception is applicable, i.e. that Tweety is not a penguin, not an ostrich etc.. What we want is something different: we want to be able to derive that Tweety flies if it <u>cannot be shown</u> that he is <u>exceptional</u>, not only if it <u>can be shown</u> that he is <u>not exceptional</u>.

What is the underlying problem? FOL has the following property: for every set of premises A and formulas p, q we have

$$A \vdash q \Rightarrow A \cup \{p\} \vdash q.$$

i.e. additional information can never invalidate old conclusions ("$\vdash$" denotes the provability relation). This property is called the monotonicity of FOL. As we saw in the Tweety example, a reasonable use of a default rule allows us to derive a default conclusion if nothing indicates an exception. If later new information is obtained which shows that the case at hand is exceptional, contrary to our expectation, then the former conclusion has to be withdrawn. The intended form of reasoning based on default rules, therefore, is nonmonotonic. Consequently a logic which intends to model such reasoning also has to be nonmonotonic.

In this paper we will mainly be interested in default reasoning. There are, however, some other forms of nonmonotonic reasoning which at least have to be mentioned briefly. As mentioned before, default reasoning allows the derivation of plausible conclusions from incomplete knowledge if there is no contradictory evidence. A very different form of nonmonotonic reasoning is autoepistemic reasoning. Here knowledge about what is known and not is used to derive correct conclusions. The standard example is

I know all my brothers.	premise
<u>I have no information as to whether Peter is my brother.</u>	derived
Peter is not my brother.	

Nonmonotonicity in this case arises because the meaning of statements about one's knowledge is context-dependent: if we get in our example additional information telling us that Peter actually is my brother, then we know that the first premise was wrong when it was used for the derivation of "Peter is not my brother". But this does not mean that the premise has to be given up. The premise refers to my knowledge. Changing the knowledge (here: including "Peter is my brother") also changes the meaning of the premise. It is reasonable to assume that the premise now, given the new information, is absolutely correct. For a more detailed analysis of autoepistemic reasoning see (Moore 85).

Also reasoning with inconsistent information is nonmonotonic. Assume you have the inconsistent premises $\{p, \neg p, q, r, s\}$. What would an intelligent agent do in such a situation? Certainly not throw away all information. One way to handle this situation is to consider maximal consistent subsets of the premises and to conclude what is true in all of them. In our example, this would have the effect of disregarding p and $\neg p$ and using only the other premises for derivations. Of course, the addition of new information, e.g. $\neg q$, may "knock-out" other formulas. Hence this type of reasoning is nonmonotonic. We will discuss inconsistency tolerant reasoning in Section 3.

Before we turn to some of the important formalizations of nonmonotonic reasoning, let us discuss some examples which demonstrate its relevance for AI. Historically, much of the work within this area has been motivated by the famous frame problem: How can we adequately represent that most things do not change when an event occurs?

The problem was first noticed in the context of the situation calculus, which was developped by McCarthy and Hayes to model reasoning about actions and events in a logical framework. In order to represent the effects of events, facts are indexed by the situations in which they hold, for instance:

HOLDS(IN(FRED, KITCHEN), SIT105)

HOLDS(COLOR(KITCHEN, RED), SIT105)

Events produce new situations:

SIT106 = RESULT(GO(FRED, BATHROOM), SIT105)

Additional axioms describe how events change the world:

$\forall x, y, s.$ HOLDS(IN(x,y), RESULT(GO(x, y), s)

Now the problem is: what is the color of the kitchen in situation Sit106? To derive that the kitchen still is red the following axiom is needed:

$\forall x, y, v, w, s.$ HOLDS(COLOR(x,y), s) -> HOLDS(COLOR(x,y), RESULT(GO(x, y), s))

Such "frame axioms" are necessary for every pair consisting of an event and a fact which remains unchanged when the event occurs. The frame axioms make the whole situation calculus approach highly impractical. As a possible solution one could use a "persistence default" like:

Events typically don't change facts unless explicitly stated otherwise.

Given an adequate representation of this default, only the effects of events have to be represented. No further frame axioms are needed any longer. Their conclusions can be obtained by the persistence default.

The frame problem has stimulated many researchers to work in the area of NMR. Meanwhile it has turned out, however, that a too simplistic representation of the persistence default in any of the existing nonmonotonic logics often does not lead to the intuitively expected results. An analysis of the arising problems together with a discussion of various proposed solutions can be found in (Hanks, McDermott 87).

A related problem is the qualification problem: How can we adequately represent that actions may fail? When, for instance, you turn the car key you would expect the car to start. But, as everybody knows, something may go wrong: the tank may be empty, the battery may be dead, someone may have stolen the engine, etc. Again, the list of possible exceptions is open-ended. Probably none of us would try to prove that nothing can go wrong before turning the car key. We base our plans on the expectation that things will behave as usual. Again, it is a natural idea to use defaults for describing the effects of events formally.

There are many more application areas where NMR is of importance. In diagnosis applications, defaults can be used to describe the normal behavior of devices. Objects which do not behave according to their default behavior are those with malfunctions. The defaults which could not be applied thus show what went wrong. In natural language understanding, we use defaults to solve ambiguities. In vision, defaults can be used to create full images from partial scenes. Also, legal reasoning is nonmonotonic. Consider the following legal rules (Gordon 87):

> *Contracts are valid.*
> *Contracts with minors are invalid.*
> *Contracts which have been ratified by a guardian of a minor are valid.*

It is easy to see that in all of these cases additional information may lead to the revision of former conclusions.

Nonmonotonic systems have quite a long tradition in AI. An early example are frame systems such as those used in many commercial expert system building tools. Frame systems allow class hierarchies and typical properties of instances of these classes to be described. Frames are class descriptions, slots represent properties of class instances and their default values. The basic idea is that, in case of conflicting defaults, the most specific information wins. Consider the following example (the language used for the frame and instance definitions is the one used in GMD's BABYLON expert system tool):

```
(DEFFRAME CAR
      (SLOTS   (SEATS 5)
               (CYLINDERS 4)
               (WHEELS 4)))
```

```
(DEFFRAME SPORTSCAR
      (SUPERS CAR)
      (SLOTS   (SEATS 2)
               (PRICE HIGH)))
```

```
(DEFINSTANCE SPEEDY OF SPORTSCAR)
```

Given these definitions we obtain that SPEEDY has 2 SEATS and 4 CYLINDERS. The value 5 for SEATS specified in the definition of CAR is not used, as CAR is a superclass of SPORTSCAR. If we extend the definition of SPORTSCAR as follows:

```
(DEFFRAME SPORTSCAR
      (SUPERS CAR)
      (SLOTS   (SEATS 2)
               (PRICE HIGH)
               (CYLINDERS 6)))
```

then we can derive that SPEEDY has 6 CYLINDERS, i.e. additional information has led to the revision of a former conclusion.

Another well-known nonmonotonic system is PROLOG with negation as failure. Consider a simple example:

FLIES(_X) :- BIRD(_X), NOT ABNORMAL(_X).
ABNORMAL(_X) :- PENGUIN(_X).
BIRD(TWEETY).

The goal FLIES(TWEETY)? yields SUCCESS given these premises, as ABNORMAL(TWEETY) cannot be proved. If we add, however, the premise

PENGUIN(TWEETY).

then the same goal is no longer derivable in PROLOG. PROLOG is nonmonotonic.

As a last example we want to mention the closed world assumption (CWA), which is well-known from the field of deductive databases. In many database applications, it is reasonable to assume completeness of the available positive information. In this case, missing information can be interpreted as negative information. For instance, if in a database of flight connections no direct connection between Bonn and London is mentioned, then this usually means that there is no such connection. Obviously, the number of facts in a database can be reduced tremendously if only positive information has to be explicitly represented.

Formally, the CWA is defined as follows:

Definition 1: *Let T be a set of formulas. p is provable from T under the CWA iff*

$$T \cup ASS(T) \vdash p$$

where ASS(T) := {¬q| q is atomic and not T ⊢ q}.

It is obvious that provability under the CWA is nonmonotonic.

These examples show that nonmonotonicity has always played an important role in AI and computer science. Early nonmonotonic systems, however, were not well-understood, or they were too restricted and it was impossible to generalize their underlying ideas in an obvious way. The CWA, for instance, may lead to inconsistency if we admit disjunctions in the premises. Without a general nonmonotonic logic it is far from clear how to extend the CWA in a reasonable way to more general cases. Nonmonotonicity, therefore, has become an important research topic of its own since about 1979. The first goal was to find an adequate formalization of NMR which was at least as general as FOL. The most important results of this research will be presented in the next section.

2. Formalizations of Nonmonotonic Reasoning

Before we present in this section some of the most important formalizations of nonmonotonic reasoning, let us briefly discuss a problem which makes a formalization technically quite difficult: the problem of conflicting defaults. Consider the following example:

Quakers (typically) are pacifists.

Republicans (typically) are not pacifists.
Nixon is a quaker and a republican.

Is Nixon a pacifist? There are two conflicting defaults, which can be used to derive different, mutually inconsistent conclusions. Note that there is nothing wrong with the defaults themselves. However, they seem to induce different sets of acceptable beliefs. As we will see, some of the formalizations generate multiple belief sets (extensions) in such a case. In each of the generated extensions, a maximal set of defaults has been applied subject to the condition that the results be consistent. Given multiple extensions, the question arises what the derivable formulas are. Two different views are possible. In the skeptical view, only what is true in all extensions is taken to be a conclusion. This view, as we will see, is implicit in McCarthy's circumscription approach. Reiter, on the other hand, follows the less cautious view that each of the possible extensions can be taken as a set of beliefs one might justifiably adopt.

The three approaches we will discuss in this section are:

Circumscription (McCarthy, Lifschitz): The notion of entailment is defined in a new way. Only a subset of the models, the minimal or preferred models, are used in the definition of entailment. Syntactically, the uninteresting models are eliminated by an additional formula schema or a second order formula,

The modal approach (McDermott & Doyle, Moore): A modal operator is used to explicitly represent whether a formula is believed or not. A fixed point operator is used to define the extensions,

The default rule approach (Reiter): Defaults are represented as nonstandard inference rules. Extensions are defined as fixed points of an operator. The operator guarantees that in each extension as many defaults have been applied as possible without destroying consistency.

We will now discuss the basic ideas underlying these approaches in more detail.

2.1 Circumscription

Circumscription allows us to minimize the extension of certain predicates. The nonmonotonic theorems of a set of premises T are defined to be the monotonic theorems of T together with some other (second order) formula or formula schema X. The role of X is to eliminate all models of T in which the extension of the predicate to be minimized is not minimal. Various versions of circumscription have been defined. We will consider here only the simplest version, predicate circumscription. Here is the definition:

Definition 2: *Let T be a first order formula (the conjunction of a finite first order theory) containing a predicate P, T(ϕ) is obtained from T by replacing all occurrences of P by ϕ. The predicate circumscription of P in T is the schema*

$$T(\phi) \ \& \ (\forall x.\phi(x) \rightarrow P(x)) \rightarrow (\forall x.P(x) \rightarrow \phi(x))$$

All instances of this schema can be used for derivations. For ϕ predicate expressions with the same arity as P can be substituted. A predicate expression with arity n is a lambda expression of the form $\lambda x_1,...,x_n.F$ where F is a formula and x_i a variable. During the substitution the open variables in F are replaced by the arguments of ϕ. Here is a simple example (McCarthy 80):

$$T = \text{ISBLOCK}(A) \ \& \ \text{ISBLOCK}(B) \ \& \ \text{ISBLOCK}(C)$$

Predicate circumscription of ISBLOCK in T yields:

$\phi(A)$ & $\phi(B)$ & $\phi(C)$ & $\forall x.(\phi(x) \rightarrow \text{ISBLOCK}(x)) \rightarrow \forall x.(\text{ISBLOCK}(x) \rightarrow \phi(x))$

Substitution of $\lambda x.(x=A \vee x=B \vee x=C)$ for ϕ yields:

$(A=A \vee A=B \vee A=C)$ & $(B=A \vee B=B \vee B=C)$ & $(C=A \vee C=B \vee C=C)$ &

$\forall x.(x=A \vee x=B \vee x=C) \rightarrow \text{ISBLOCK}(x))$

$\rightarrow$

$\forall x.(\text{ISBLOCK}(x) \rightarrow (x=A \vee x=B \vee x=C))$

The antecedent of this implication is derivable from T and we can derive that A, B, C are the only blocks. It is not difficult to see that whenever the original set of premises changes (e.g. when ISBLOCK(D) is added as another conjunct), the schema also changes and we need different substitutions to get an instance with true antecedent. This is what makes circumscription nonmonotonic.

As a general principle for default reasoning McCarthy, proposes to use abnormality predicates. A default like the one used in the flying birds example is represented in the following way:

$\forall x.\text{BIRD}(x)$ & $\neg \text{AB1}(x) \rightarrow \text{FLIES}(x)$

For every default, a specific AB-predicate is introduced representing the objects which behave abnormally with respect to this default. All AB-predicates are circumscribed; that is, as few objects as possible are expected to be abnormal. Unfortunately, it turns out that more complicated versions of circumscription are needed to produce intuitively satisfying results. Moreover, the second order version of circumscription, where the schema is replaced by a second order formula, has theoretical advantages. For a detailed overview see (Genesereth, Nilsson 87).

2.2 Autoepistemic Logic

Moore's autoepistemic logic (AEL) (Moore 85) is the most prominent example of the modal approach. Moore tries to model an ideal introspective agent able to reason about his own beliefs and disbeliefs; that is, the agent knows exactly what he knows and what he does not know. For this purpose, a modal operator L is introduced: Lp stands for "it is believed that p". Our bird rule is represented as the schema

$\text{BIRD}(x)$ & $\neg L \neg \text{FLIES}(x) \rightarrow \text{FLIES}(x)$

The question now is: what are the sets of beliefs a rational agent may adopt given complete insight into his own knowledge. Moore defines these sets of beliefs, the extensions of a set of premises A, as follows (again "⊢" stands for classical derivability, formulas of the form Lp are treated as propositional constants):

Definition 3: *Let A be a set of premises. T is an extension of A iff*

$$T = \{p \mid A \cup Bel(T) \cup Disbel(T) \mid\text{-} p\}, \text{ where}$$

$$Bel(T) = \{Lq \mid q \in T\} \text{ and } Disbel(T) = \{\neg Lq \mid q \notin T\}.$$

Extensions hence are sets of formulas which contain the premises, which are deductively closed, and which contain Lp iff they contain p and which contain ¬Lp iff they do not contain p. For instance, the theory A =

BIRD(TWEETY) & ¬L¬FLIES(TWETY) -> FLIES(TWEETY)

BIRD(TWEETY)

has exactly one extension. As ¬FLIES(TWEETY) is underivable, even if arbitrary formulas of the form Lp or ¬Lp are added to A, ¬L¬FLIES(TWEETY) must be contained in the extension. Since this formula and the second premise is contained in the extension, FLIES(TWEETY) must also be contained in it, otherwise the extension would not be deductively closed.

Generally there may be more than one extension. For instance, in the Nixon example discussed in the beginning of this section, two extensions are generated corresponding to the different belief sets supported by the conflicting defaults.

It has recently been shown that AEL and the logic which will be presented next, default logic, are in a certain sense "equivalent" (Konolige 88). AEL theories can be translated into default logic theories such that the default logic extensions are the modal-operator-free part of some (arguable the most interesting) AEL extensions. We, therefore, omit further examples here and directly present default logic.

2.3 Default Logic

In default logic (DL) (Reiter 80) defaults are represented as nonstandard inference rules. A default theory is a pair (D,W) where W is a set of standard first order formulas representing the certain facts. D is a set of defaults of the form

$$\frac{A:\ B1,...,Bn}{C}$$

where A, B1, ..., Bn and C are classical formulas. The intuitive reading of the default is: if A is derivable, ¬B1, ..., ¬Bn is not derivable, then derive C. Again a fixed point definition is needed to define the extensions of a default theory:

Definition 4: *Let (D,W) be a default theory and S a set of formulas. $\Gamma(S)$ is the smallest set such that*

1) $W \subseteq \Gamma(S)$,

2) $\Gamma(S)$ *is deductively closed,*

3)   *if* $\frac{A:\ B1,...,Bn}{C} \in D, A \in \Gamma(S), \neg Bi \notin S$ *(for all i), then* $C \in \Gamma(S)$.

Fixed points of Γ are called extensions of (D,W). Intuitively, the extensions contain the certain knowledge from W, are deductively closed and contain the conclusions of as many defaults as possible. Here are some simple examples:

D	W	fixed point(s)
$\dfrac{Bird(x):Flies(x)}{Flies(x)}$	Bird(Tw)	Th(W $\cup$ {Flies(Tw)})
$\dfrac{Bird(x):Flies(x)}{Flies(x)}$	Bird(Tw) Peng(Tw) $\forall x.Peng(x) \rightarrow \neg Flies(x)$	Th(W)
$\dfrac{Bird(x):Flies(x)}{Flies(x)}$ $\dfrac{Peng(x):\neg Flies(x)}{\neg Flies(x)}$	Bird(Tw) Peng(Tw)	Th(W $\cup$ {Flies(Tw)}) Th(W $\cup$ {$\neg$ Flies(Tw)})
$\dfrac{Bird(x):Flies(x) \,\&\neg Peng(x)}{Flies(x)}$ $\dfrac{Peng(x):\neg Flies(x)}{\neg Flies(x)}$	Bird(Tw) Peng(Tw)	Th(W $\cup$ {$\neg$ Flies(Tw)})

Figure 1: Default theories and their extensions

It is also possible that there is no extension at all. Consider the default theory where W is empty and D consists of the single default

$$\frac{true: \neg A}{A}$$

We can show that no set S is a fixed point. Assume S does not contain A. Then $\Gamma(S)$ contains A as the default can be applied and hence S is no fixed point. Now assume S contains A. $\Gamma(S)$ does not contain A as the default becomes inapplicable. Again S is no fixed point. The non-existence of extensions is usually treated as inconsistency.

(Brewka 91) describes a modification of default logic which produces more intuitive results and behaves more regular than Reiter's original logic in some cases.

3. Preferred Subtheories: A Framework for Default Reasoning

We saw in the introduction that the "standard" approaches to formalizing nonmonotonic and, in particular, default reasoning start from a consistent set of premises (otherwise no interesting result at all is obtained) and extend the inference relation to get more than just the classically derivable formulas. Here we will present an approach based on an alternative view. What makes a default a default? What distinguishes it from a fact? Certainly our attitude towards it in case of a conflict, i.e. an inconsistency. If we take this view seriously, then the idea of default reasoning as a special kind of inconsistency handling seems quite natural. There need not be a problem with inconsistent premises as long as we provide ways of handling the inconsistency adequately (in other words, if we modify the inference relation so that in case of inconsistency fewer, i.e. not all formulas are derivable).

In this section we will first present a simple general framework for defining nonmonotonic systems based on this view. We then show that Poole's approach to default reasoning (Poole 88) is a simple instance of this framework and discuss a limitation of his approach, due to the inability of representing priorities between defaults. Sect. 2.3 presents a generalization of Poole's approach, where several layers of possible hypotheses, representing different degrees of reliability, are possible. A further generalization, based on a partial ordering between premises, is described in Sect. 2.4. In both approaches, a formula is provable from a theory if it is possible to construct a consistent argument for it based on the most reliable hypotheses.

3.1   A Framework for Nonmonotonic Systems

A standard way of handling inconsistent premise sets uses maximal consistent subsets of the premises at hand. Since, in general, there is more than one such maximal consistent subset, provability is defined as provability in all such sets. The idea behind the "maximal" is clear: we want to modify the available information as little as possible. The notion of maximal consistent subsets per se, however, does not allow us to express, say, that *Tweety flies* should be given up instead of *Tweety is a penguin*, if we know that penguins don't fly. To be able to express such preferences, we have to consider not all maximal consistent subsets, but only some of them, the preferred maximal consistent subsets, or simply: *preferred subtheories*. This idea is similar to preferential entailment where not all, but only a subset of the models is taken into account.

The notion of a preferred maximal consistent subset is not new: it dates back to (Rescher 64). Rescher defined a particular ordering of subtheories for hypothetical reasoning. He did not, however, apply this idea to default reasoning.

We are now in a position to define a weak and a strong notion of provability:

Definition 5: *A formula p is weakly provable from a set of premises T iff there is a preferred subtheory S of T such that S |- p.*

Definition 6: *A formula p is strongly provable from a set of premises T iff for all preferred subtheories S of T we have S |- p.*

These notions, roughly, correspond to containment in at least one or in all extensions in the fixed point approaches to default reasoning. In fact, we can also introduce the notion of an extension in the following way:

Definition 7: *E is an extension of a set of premises T iff there is a preferred subtheory S of T and E = Th(S).*

To specify exactly what the preferred subtheories are, we will impose in the rest of the paper a certain structure on the premises T. In one approach (Section 2.2), for instance, we will split T into several levels $T_1, ..., T_n$. This additional structure will be used to define the preferred subtheories of T. For sake of simplicity, we will also speak of preferred subtheories (and extensions) of these structures and leave the premise set T implicit.

One important aspect of this approach should be noted: the provable formulas of our theories depend on the syntactic form of the premises. It may make an important difference whether, for instance, a set of premises contains both A and B, or the logically equivalent single formula A & B. Assume there is a preferred subtheory S inconsistent with A, but not with B. If both A and B are given, then B must be contained in S. If, however, our premise is A & B, then this need not be the case.

We could avoid this by introducing a certain normal form for formulas. However, we don't see this unusual behaviour as a drawback at all. It increases expressiveness. We can express that two formulas A and B should be accepted or given up together. It makes perfect sense to distinguish between situations where A as well as B are possible, unrelated hypotheses and those where A & B is one hypothesis.

Consider the following example: someone tells you that Michael was sitting in a bar drinking beer. Call "Michael was in the bar" A and "Michael was drinking beer" B. Now someone else gives you the more reliable information "Michael was working at the university". Call this information C. Clearly, C is inconsistent with A, but not with B. However, it seems reasonable to give up B also in this case. In our approach, this can be achieved by using the premise A & B. Note that replacing a single premise P by an equivalent single premise Q does not change the resulting preferred subtheories.

What we have so far is just a general framework. To obtain a specific instance of the framework, i.e. a real nonmonotonic system, the preference relation on subtheories must be defined. We will first show that Poole's approach can be seen as such an instance.

3.2   Poole's System: Default Reasoning as Theory Construction

David Poole (Poole 88) recently presented a simple and elegant, yet quite expressive approach to default reasoning. In Poole's framework the user provides

1)   a consistent set F of closed formulas, the facts about the world,

2)   a set Δ of, possibly open, formulas, the possible hypotheses.

Definition 8 (Poole): *A scenario of F and Δ is a set D ∪ F where D is a set of ground instances of elements of Δ such that D ∪ F is consistent.*

Definition 9 (Poole): *A formula g is explainable from F and Δ iff there is a scenario of F and Δ which implies g.*

Definition 10 (Poole): *An extension of F and Δ is the set of logical consequences of a (set inclusion) maximal scenario of F and Δ.*

The terminology reflects that Poole uses his framework not only for prediction, but also for explanation of observed facts. Since our main interest here is default reasoning, we will also call Δ the set of defaults. Poole has shown that many of the standard examples involving defaults can be handled adequately with this simple framework.(Poole 88). He uses the following naming technique: for a hypothesis $w(x) \in \Delta$ with free variables x, a new predicate symbol p_w of the same arity is introduced. Poole shows that $w(x)$ can equivalently be replaced by $p_w(x)$, if the formula

$$\forall x.p_w(x) \rightarrow w(x)$$

is added to F. Poole uses the notation $p_w(x):w(x)$ as an abbreviation for this case. Thus, the expressiveness of the system is not restricted if only atoms are admitted in Δ.

The use of names allows the applicability of a default to be blocked when needed. If we want a default $p_w(x)$ to be inapplicable in situation s we simply have to add $\forall x.s \rightarrow \neg p_w(x)$ to our facts. (Poole 88) contains many nice examples of how this technique can be used.

It is not difficult to see that Poole's approach can be obtained as a simple instance of our preferred subtheory framework: if we define the preferred subtheories of $\Delta' \cup F$ (Δ' is obtained from Δ by replacing open formulas with all of their ground instances) as those subtheories containing F, then weak provability and Poole's explainability coincide.

Poole's approach is simple and elegant, and its expressiveness is astonishing. Moreover, a quite efficient Prolog-implementation exists (Poole et al. 86). However, as shown in (Brewka 89), there is no simple way of representing priorities between defaults: This motivated the generalizations presented in the rest of this section.

3.3. First Generalization: Levels of Reliability

The following figure illustrates the basic idea of Poole's approach: we have two levels of theories, the basic level can be seen as premises which must be true (and consistent); the second level is a level of hypotheses which are less reliable.

Δ	Hypotheses
F	Facts

We generalize these ideas in two respects. Firstly, we do not require the most reliable formulas (i.e. T1) to be consistent. In our generalization every formula is in principle defeasible. And secondly, we allow more than just two levels. This can be illustrated by the following graphic:

Tn	Hypotheses
Tn-1	Hypotheses
	...
T2	Hypotheses
T1	Hypotheses

The idea is that the different levels of a theory represent different degrees of reliability. The innermost level is the most reliable one. If inconsistencies arise, the more reliable information is preferred. Intuitively, a formula is provable if we can construct an argument for it from the most reliable available information. Of course, there may be conflicting information with the same reliability. In this case we get multiple extensions, i.e. two contradicting formulas can be provable in the weak sense. The fact that there are no in principle unrefutable "premises" makes it possible to treat all levels uniformly. For instance, we can add to any theory information which is even more reliable than the current innermost level.

We now show how these intuitive ideas can be made precise using the preferred subtheory approach.

Definition 11: *A default theory T is a tuple $(T_1, ..., T_n)$, where each T_i is a set of classical first order formulas.*

Intuitively, information in T_i is more reliable than that in T_j if i<j. A default like *birds fly* can be represented as the set of all ground instances of a schema BIRD(x) -> FLIES(x). For sake of simplicity we will write T_i = {..., P(x), ...} if we want to express that T_i contains all ground instances of P(x). Note again the important difference between universally quantified formulas and schemata containing free variables.

The preferred subtheories remain to be defined:

Definition 12: *Let $T=(T_1,...,T_n)$ be a default theory. $S = S_1 \cup ... \cup S_n$ is a preferred subtheory of T iff for all k $(1 \leq k \leq n)$ $S_1 \cup ... \cup S_k$ is a maximal consistent subset of $T_1 \cup ... \cup T_k$.*

In other words, to obtain a preferred subtheory of T, we have to start with any maximal consistent subset of T_1, add as many formulas from T_2 as consistently can be added (in any possible way), and continue this process for T_3, ..., T_n.

The following simple examples show how the different levels can be used to express priorities between defaults:

T1 = {BIRD(TWEETY), $\forall$x.PENGUIN(x) -> ¬FLIES(x)}

T2 = {BIRD(x) -> FLIES(x)}

FLIES(TWEETY) is strongly provable.

T1 = {BIRD(TWEETY), $\forall$x.PENGUIN(x) -> ¬FLIES(x), PENGUIN(TWEETY)}

T2 = {BIRD(x) -> FLIES(x)}

¬FLIES(TWEETY) is strongly provable. This example also illustrates the importance of the distinction between schemata and universally quantified formulas. If we wouldn't use a schema in T2 but instead a quantified formula, then this formula wouldn't be usable if there is a single nonflying bird.

If there is a penguin who does fly, then we can use the following representation where *penguins don't fly* is given higher priority than *birds fly*:

T1 = {BIRD(TWEETY), PENGUIN(TWEETY), PENGUIN(TIM), FLIES(TIM)}

T2 = {PENGUIN(x) -> ¬FLIES(x)}

T3 = {BIRD(x) -> FLIES(x)}

3.4   Second Generalization: Partially Ordered Defaults

For many problems, the introduction of levels of reliability as described above is sufficient to express the necessary priorities between defaults. Sometimes, however, we want to leave open whether a formula p is more or less reliable than another formula q. Here is an example about how certain political groups in Germany would deal with unemployment. The social democrats (SPD) usually believe that there are not enough jobs and that we should simply share the work available, i.e. everybody should work less. However, reducing everybody's salary would be bad for economy. A subgroup of the social democrats following Lafontaine, however, agree with the first, but not with the second position since, according to their view, working less without reducing salaries would make work too expensive. According to employers there is enough work. They believe that salaries are simply too high. These ideas could be formalized by the following defaults:

(1)   SPD -> LESS-WORK & ¬LESS-MONEY

(2)   LAF -> LESS-WORK & LESS-MONEY

(3)   EMPLOYER -> ¬LESS-WORK & LESS-MONEY

LAF is a subclass of SPD, i.e. information about Lafontaine-supporters is more specific than information about social democrats in general. We certainly want to give (2) priority over (1) in this case. But how about (3)? The approach from the last section forces us to choose exactly one level for each formula, i.e. to specify a priority either between (1) and (3) or between (2) and (3). There seems to be no reason why we should want this in the example.

This problem can be avoided if we allow the degrees of reliability to be represented by an arbitrary (strict) partial ordering of the premises, instead of different levels. A default theory now is a pair (T, <), where the relation < is defined over T, and p < q, intuitively, states that p is more reliable than q. Again we have to define the preferred subtheories to obtain weak and strong provability based on such a partial ordering:

Definition 13: *Let < be a strict partial ordering on a (finite) set of premises T. S is a preferred subtheory of a default theory (T, <) iff there exists a strict total ordering $(t_1, t_2, ... , t_n)$ of T respecting < (i.e. $t_j < t_k$ => j < k) such that $S=S_n$ with*

   $S_0 := \{\}$, and for $0 \leq i < n$

$S_{i+1} := $ *if* t_{i+1} *consistent with* S_i *then* $S_i \cup \{t_{i+1}\}$ *else* S_i.

The generalization to the infinite case is straightforward if we forbid infinite descending chains in the orderings.

If we define in our above example the ordering to be (2) < (1), then LESS-WORK & LESS-MONEY is strongly provable from SPD and LAF ("from some formulas" here means that these formulas are smaller than (1), (2) and (3) with respect to <). From SPD and EMPLOYER both LESS-WORK & ¬LESS-MONEY and ¬LESS-WORK & LESS-MONEY are weakly provable. The two corresponding preferred subtheories are generated from different total orderings respecting <. Similarly, we obtain two preferred subtheories given LAF and EMPLOYER. In this case LESS-WORK & LESS-MONEY is provable from one preferred subtheory.

The example shows that we were able to express the wanted preference between (2) and (1) without having to introduce any unwanted preference involving (3).

There is obviously a common intuition behind our approach, in particular the last generalization, and the notion of epistemic entrenchment used in work on theory revision (Gärdenfors, Makinson 88) and (Gärdenfors 88). These authors, however, are interested in belief sets, i.e. deductively closed sets of formulas, and the changes of such states when new information is obtained. They are not interested in deriving plausible conclusions from, possibly inconsistent, premises. Therefore epistemic entrenchment is defined for the whole logical language, whereas for our purposes it is sufficient to order premises. Moreover, they require the new knowledge state after the addition of new information to be uniquely determined by the epistemic entrenchment, whereas we allow multiple preferred subtheories whenever no priority between conflicting defaults is specified.

See (Brewka 90) for a discussion of other related work and more motivating examples for both generalizations.

In this section we presented two generalizations of Poole's approach to default reasoning as particular instances of a general nonmonotonic framework. The first generalization extends his original approach in two respects: 1) we allow several levels of reliability instead of only two and 2) treat all levels uniformly, i.e. there are no unrefutable premises. The second generalization introduces a partial ordering on the premises instead of the levels. In Poole's theory the applicability of a default can be blocked, but there is no way of representing priorities between defaults in the sense that one of two conflicting defaults is not applied if the other one can be applied. Our systems provide a natural means of representing such priorities.

There is always a tradeoff between expressiveness and simplicity. Obviously, we had to give up some of the simplicity and elegance of Poole's system in order to increase expressiveness and to allow for the representation of default priorities. But still we are much closer to classical logic than many other systems: we don't need modal operators, nonstandard inference rules, fixed point constructions, second order logic or abnormality predicates. This should make it simpler to integrate default reasoning with other forms of commonsense reasoning.

The main advantage of this approach to default reasoning, as we see it, is that the problem of handling inconsistent information - a problem every commonsense reasoner has to deal with anyway - is implicitly solved.

4. How to implement a nonmonotonic reasoning system?

All approaches examined in the last two sections are not semi-decidable, i.e.one can prove that no correct and complete proof procedure exists for the general case. But this does not mean that these approaches are entirely unimplementable. Three solutions are possible:

1) We can restrict the logics and try to find computable subsets.

2) We can look at some of the existing nonmonotonic programs and see whether they can be interpreted in terms of the some nonmonotonic logic.

3) We can try to approximate the desired results.

A number of restrictions of the logics have been investigated, in particular in the context of circumscription. It has been shown that in many cases circumscription can be reduced to first order reasoning (Lifschitz 85) or to logic programming (Gelfond, Lifschitz 89). These important results show that often standard theorem proving techniques can be used for NMR. Proof procedures for subsets of AEL and DL have also been described (Reiter 80; Brewka 90; Junker, Konolige 90). Poole and his colleagues have presented a PROLOG- implementation of his approach (Poole et al. 86).

The main focus of the second approach have been nonmonotonic inheritance systems, which allow a hierarchy of objects and object classes to be represented in which properties are inherited by more specific classes. A special class of such systems, frame systems, were discussed in Section 1. It has been shown that some of these systems can be interpreted as specialized nonmonotonic provers. An example is the analysis of frame systems in terms of circumscription in (Brewka 87).

Regarding the third approach, a number of implemented systems, such as WATSON (Goodwin 87) or CAPRI (Freitag, Reinfrank 88) approximate the theoretically expected results in the following way: a problem solver operates on a knowledge base consisting of inference rules of the form

$$\text{IF } B_1,...,B_n \text{ THEN C UNLESS E1,...,Em}$$

Such a rule roughly corresponds to the Reiter default rule

$$\frac{B1 \ \& \ ... \ \& \ Bn: \neg E1, ..., \neg Em}{C}$$

The rule can be applied if all its IF-conditions are proven and its UNLESS-conditions are *unproven in the current problem solving state* (as opposed to *unprovable*). If an UNLESS-condition is proven later, then the conclusion is withdrawn. A truth maintenance system (TMS) records all current beliefs and their dependencies. Whenever a rule is applied, a justification for its conclusion is generated and integrated into the dependency network. The TMS then updates its dependency network and computes an acceptable belief set based on the rules fired so far. This is done by labeling the nodes in the network with IN or OUT depending on whether they belong to the belief set or not. Thus an "extension" of the problem solver's knowledge base is incrementally approximated.

Here is an example. Assume the following rules have been applied

BIRD-TW

IF BIRD-TW THEN FLIES-TW UNLESS NOT-FLIES-TW
IF PENGUIN-TW THEN NOT-FLIES-TW

In this case the following dependency network is created (the graphical notation is taken from (Goodwin 87). Gates represent justifications. The input of a gate corresponds to the justification's IF-conditions or, if there is a black point at the gate, to its UNLESS-conditions. The output node corresponds to a rule's conclusion):

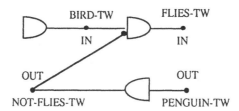

The IN/OUT labeling of nodes has been computed by the TMS in such a way that:

- a node is IN iff it has a valid justification, i.e. one where all nodes corresponding to IF-conditions are IN and all nodes corresponding to UNLESS-conditions OUT.

- no node (directly or indirectly) justifies itself.

If in the next step the premise rule PENG-TW is applied (that is a rule with no IF and UNLESS conditions), we get the following updated net:

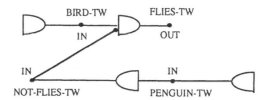

It should be noted that the TMS does not know anything about the meaning of the nodes. It uses only the known dependencies between nodes (we simplified our presentation somewhat and used the formulas from the rules as names for the nodes, real TMS's invent their own names and don't care about the formulas). There is one exception, however: False-nodes represent inconsistent states. If the TMS has produced a labeling with an IN false-node, it tries to remove the inconsistency by adding a self-generated justification. Here is a slight modification of our former example:

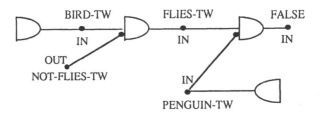

In this case the addition of a justification corresponding to the rule

IF BIRD-TW, PENGUIN-TW THEN NOT-FLIES-TW

removes the inconsistency. This process is called dependency directed backtracking. Various algorithms to generate justifications to restore consistency have been proposed in the literature (Doyle 79) (Petrie 87).

The various approaches and topics discussed in this paper show that the problem of handling incomplete knowledge has been widely recognized in Artificial Intelligence. A great number of interesting approaches have been proposed, their properties and the relation between them has been studied extensively. Moreover, a number of implementations exist. However, a real validation of any of these approaches in realistic applications is still lacking. Further progress seems possible only if the formalizations are tested in real problem domains. More efficient implementations are necessary for this purpose, and larger knowledge bases should be built with these systems. Good old Tweety is simply not enough for the next decade of research in this area.

Acknowledgments

I would like to thank Tom Gordon for helpful comments and for polishing my English.

References

(Brewka 87) Brewka, Gerhard: The Logic of Inheritance in Frame Systems, Proc. IJCAI 87, 1987

(Brewka 90) Brewka, Gerhard: Nonmonotonic Reasoning - Logical Foundations of Commonsense, Cambridge Tracts in Theoretical Computer Science 12, Cambridge University Press, 1990

(Brewka 91) Brewka, Gerhard: Cumulative Default Logic - In Defense of Nonmonotonic Inference Rules, Artificial Intelligence, to appear

(Doyle 79) Doyle, J.: A Truth Maintenance System, Artificial Intelligence 12, 1979

(Freitag, Reinfrank 88) Freitag, H., Reinfrank, M.: A Non-Monotonic Deduction System Based on (A)TMS, Proc. ECAI 88, München, 1988

(Gärdenfors 88) Gärdenfors, Peter: Knowledge in Flux, MIT Press, Cambridge, MA, 1988

(Gärdenfors, Makinson 88) Gärdenfors, Peter, Makinson, David: Revisions of Knowledge Systems Using Epistemic Entrenchment. In: Vardi, M. (ed): Proceedings of the Second Conference on Theoretical Aspects of Reasoning about Knowledge, Morgan Kaufmann, Los Altos, 1988

(Gelfond, Lifschitz 88) Gelfond, Michael, Lifschitz, Vladimir: Compiling Circumscriptive Theories into Logic Programs, Proc. 2nd Int. Workshop on Nonmonotonic Reasoning, Springer, LNCS 346, 1988

(Goodwin 87) Goodwin, James W.: A Theory and System for Non-Monotonic Reasoning, Linköping University, Computer and Information Science Dep., Dissertation No. 165, 1987

(Gordon 87), Gordon, Thomas F.: Oblog 2 - A Hybrid Knowledge Representation System for Defeasible Reasoning, Proc. 1st Intl. Conference on Artificial Intelligence and Law, Boston, ACM Press, 1987

(Hanks, McDermott 87) Hanks, Steven, McDermott, Drew: Nonmonotonic Logic and Temporal Projection, Artificial Intelligence 33, 1987

(Junker, Konolige 90) Junker, Ulrich, Konolige, Kurt: Computing the Extensions of Autoepistemic and Default Logic with a TMS, Proc. AAAI 90, 1990

(Konolige 88) Konolige, Kurt: On the Relation Between Default and Autoepistemic Logic, Artificial Intelligence 35 (3), 1988

(Lifschitz 85) Lifschitz, Vladimir: Computing Circumscription, IJCAI 85, 1985

(McCarthy 80) McCarthy, John: Circumscription - A Form of Nonmonotonic Reasoning, Artificial Intelligence 13, 1980

(McCarthy 84) McCarthy, John: Applications of Circumscription to Formalizing Common Sense Knowledge, Proc. AAAI-Workshop Non-Monotonic Reasoning, 1984 (auch in Artificial Intelligence 28, 1986)

(Moore 85) Moore, Robert C.: Semantical Considerations on Nonmonotonic Logic, Artificial Intelligence 25, 1985 (Kurzfassung in Proc. IJCAI 83)

(Petrie 87) Petrie, C.J.: Revised Dependency-Directed Backtracking for Default Reasoning, *Proc. AAAI 87*, 1987

(Poole 88) Poole, D.: A Logical Framework for Default Reasoning, Artificial Intelligence 36, 1988

(Poole et al. 86) Poole, D., Goebel, R., Aleliunas, R.: A Logical Reasoning System for Defaults and Diagnosis, University of Waterloo, Dep. of Computer Science, Research Rep. CS-86-06, 1986

(Reiter 80) Reiter, Raymond: A Logic for Default Reasoning, Artificial Intelligence 13, 1980

(Rescher 64) Rescher, Nicholas: *Hypothetical Reasoning*, North-Holland Publ., Amsterdam. 1964

Negation in Logic Programming:
A Formalization in Constructive Logic

François Bry

ECRC, Arabellastraße 17, D - 8000 München 81, Germany
fb@ecrc.de

ABSTRACT   *The conventional formalization of logic programming in classical logic explains very convincingly the basic principles of this programming style. However, it gives no easy or intuitive explanations for the treatment of negation. Logic Programming handles negation through the so-called "Negation as Failure" inference principle which is rather unconventional from the viewpoint of classical logic. Despite its nonclassical nature, this inference principle cannot be avoided in practice. The appropriate application of Negation as Failure requires either syntactical restrictions, or significant changes in the semantics of logic programs. In this article, we defend the thesis that these syntactical restrictions or semantical changes are naturally and simply explained by observing that logic programming in fact implements no more than a constructive fragment of classical logic. Relying on a "Conditional Fixpoint Procedure", we first define a monotonic inference procedure for logic programs with negation that are consistent in this constructive fragment. Then we show how this procedure can be extended into a "Ternary Fixpoint Procedure" for general programs. This fixpoint procedure defines a ternary logic semantics for syntactically unrestricted logic programs. Finally, we argue that the constructive interpretation of logic programming also gives a simple and natural explanation of meta-programming. Relying on this view of meta-programming, we specify different forms of reflective reasoning, in particular default reasoning.*

1. Introduction

Since *Prolog* (*Prog*ramming in *Log*ic) has been developed by A. Colmerauer and R. Kowalski, the concept of "logic programming" refers to a certain type of declarative programming languages. Logic programs are formalized as relationship between terms in a logical theory [vEK76, Kow79]. This view gives a very convincing explanation of the basic principles of declarative programming languages, which is independent from any inference principle. However, this formalization proves to be inappropriate as soon as additional features – language constructs or syntactical restrictions – are considered, that are necessary for practical applications of declarative languages.

Negation is such a construct. Its processing, which could be considered rather unlogical by mathematicians, has led either to restrict the syntax of logic programs, or to formalize their semantics in ternary logics. In this paper, we show that these syntactical restrictions and nonclassical semantics are naturally and intuitively explained by referring to a rather simple fragment of classical logic.

This restricted logic is in fact the original form of mathematical logic. It is called "constructive logic" because it rejects existence proofs that do not explicitly construct the mathematical object, the existence of which is proven. Before Cantor, the restriction to constructive proofs was almost taken for granted. By applying its diagonalization principle, Cantor has shown that an additional number can be defined from any enumerable set of real numbers. Cantor's definition is however not a construction. Therefore his proof has been first strongly criticized. The article [Cal79] provides an easy introduction to this part of mathematics and to its history. Although nonconstructive proofs are nowadays no longer criticized, constructive reasoning has survived in mathematics. Various constructive logics have been developed aside the main stream of conventional mathematics. We do not need to refer to these theories, whose notations and vocabularies are often discouraging. We shall only make use of the original and quite intuitive notion of constructive proof.

The formalization of logic programming in constructive logic is not only useful for investigating negation. It gives also a very natural explanation of meta-programming. From the constructive viewpoint, it is rather natural to define theories or programs, whose variables range over formulas. Since in constructive logic formulas represent themselves but no platonistic objects, it is possible to interpret meta-programs as first-order, constructive theories. This constructive view of meta-programming is especially useful for axiomatizing non-classical logics such as, for example, autoepistemic or default logics. Such axiomatizations can also be used as implementations.

Some authors have already observed the link between logic programming and constructive reasoning. G. Huet suggests in [Hue86] to view Prolog interpreters as proof synthetisers of a sequent calculus – a kind of constructive logic. D. Bojadziev notices similarities between Prolog and constructive reasoning in the short article [Boj86]. Constructive aspects of Prolog are investigated by C. Beckstein in relationship with *Truth Maintenance Systems* in [Bec88]. J.-Y. Girard, Y. Lafont, and P. Taylor suggest in [GLT89] a possible relationship between logic programming and linear logic [Gir87] – a constructive logic recently proposed by J.-Y. Girard. In [Bry89a] the view of logic programming as a constructive fragment of classical logic was used, for the first time as far as we know, to intuitively motivating syntactical restrictions like stratification or ternary semantics. We do not know about any work relying on constructive logic for investigating meta-programming.

This article consists of seven sections, the first of which is this introduction. In Section 2, we recall the usual formalization of logic programming without negation. Then, we consider in Section 3 the processing of negation and the difficulties resulting from it. We define a "Conditional Fixpoint Operator" in Section 4. We show how to use this operator for monotonically generating consequences from programs with negation that are constructively consistent. We show how syntactical properties like stratification [ABW88, Van88] or local stratification [Prz88] are easily motivated in constructive logic: They are necessary conditions for constructive consistency. In Section 5, we propose a ternary valued semantics for syntactically unrestricted logic programs. We define a "Ternary Fixpoint Procedure" for this semantics. Finally, we propose an interpretation of meta-programming in constructive logic in Section 6. As examples of this interpretation, we show how reflective inference principles like modal and default logics can be specified as meta-programs. Section 7 is a conclusion.

This article is translated and adapted from the text of a talk given at the 1st Workshop Information Systems and Artificial Intelligence [Bry90].

2. Logic Programming without Negation

The inference principles of logic programming are *modus ponens* and the specialization rule. By *modus ponens,* a formula G is derived from two formulas F and $F \Rightarrow G$. If c is a constant in the universe under consideration, specialization gives rise to deriving F[c] from a universal formula $\forall x \, F[x]$. These two inference rules are combined in the resolution principle [Rob65a], the knowledge of which is not necessary for our purposes.

The two inference principles *modus ponens* and specialization can be more or less directly implemented. The simplest approach consists in applying first the specialization rule for eliminating the universal quantifiers. A formula $\forall xyz \, p(x, y) \land q(y, z) \Rightarrow r(x, z)$ – usually written as $r(x, z) \leftarrow p(x, y) \land q(y, z)$ in logic programming – is for example specialized by instantiating its variables in all possible ways over the universe under consideration. This approach which applies the two inference rules specialization and *modus ponens* separately and successively is usually retained in *Truth Maintenance Systems* [Doy79, dK86]. A major drawback is the number of useless instantiated formulas that are, in most cases, generated.

A significant improvement consists in relying on the premise of the clauses – i.e., on $p(x, y) \land q(y, z)$ in the above-mentioned example – for restricting specialization: Only those instances of the clause $r(x, z) \leftarrow p(x, y) \land q(y, z)$ are needed that make the conjunction $p(x, y) \land q(y, z)$ true. Note that no subsequent instanciations are necessary if all variables appear in the premises, i.e., if the clause is *range-restricted*. It is common to describe this improvement for range-restricted clauses as a special type of resolution, *unit hyperresolution* [Rob65b, vEK76]. This approach is called *forward* or *bottom-up* reasoning, for it somehow follows the direction of the implication sign in the formulas or in the clauses. A remarkable efficiency can be achieved in this way, which applies specialization no more than necessary.

Following van Emdem and Kowalski [vEK76], the denotational semantics of a logic program is traditionally defined in terms of bottom-up reasoning. The semantics of a logic program without negation P is defined in terms of the facts that are derivable from P by iterated applications of bottom-up reasoning. A facts which is derivable from a specialization of a clause in P by one application of *modus ponens* is called a *direct consequence* of P. The set of the direct consequences of P is noted $T(P)$.

$T\uparrow^n(P)$ denotes the set of facts that are derivable from instances of clauses in P by at most n applications of the *modus ponens* rule. This set is inductively defined as follows:

$$
\begin{aligned}
T\uparrow^0 (P) &= P \\
T\uparrow^{n+1}(P) &= T(T\uparrow^n(P)) \cup T\uparrow^n(P)
\end{aligned}
$$

The sets of all facts that are derivable from P is:

$$\bigcup_{n\in N} T\uparrow^n(P) = T\uparrow^0 (P) \cup T\uparrow^1 (P) \cup ... \cup T\uparrow^n (P) \cup ...$$

It is denoted $T\uparrow^\omega (P)$. This set is the fixpoint of the function $T\uparrow$ on the program P.

The function $T\uparrow$ and the computation of its fixpoints are traditionally used for formalizing the semantics of logic programs without negation. Despite of an elegant analogy between model theory in logic and denotational semantics of programs [vEK76], it is worth to notice that the computation of a fixpoint $T\uparrow^\omega(P)$ is in principle a forward reasoning procedure.

This procedure has two forms, often called *naïve* and *semi-naïve methods*. The naïve method computes the sets $T\uparrow^n(P)$ for increasing values of n. The semi-naïve methods improves the naïve methods by only computing thoses facts of $T\uparrow^{n+1}(P)$ that have at least one premise in

$$T\uparrow^n(P) \setminus T\uparrow^{n-1}(P)$$

This aims at restricting the computation of $T\uparrow^{n+1}(P)$ as much as possible to that of

$$T\uparrow^{n+1}(P) \setminus T\uparrow^n(P)$$

The operators T and $T\uparrow$ are monotonic on programs without negation: If $P_1 \subseteq P_2$, then $T(P_1) \subseteq T(P_2)$ and $T\uparrow(P_1) \subseteq T\uparrow(P_2)$.

In order to avoid the generation of useless facts and to restrict as much as possible the generation to relevant facts, so-called backward or *top-down* reasoning methods are applied. It is a quite intuitive fact, that forward reasoning cannot take into account the constants occurring in queries. This inference technique indeed considers the queries at the very end. As opposed backward reasoning methods start from the queries. They can therefore propagate the constants occurring in queries.

Backward reasoning is close to the so-called problem reduction reasoning paradigm. For example, in order to prove a fact p(a), a backward reasoning methods searches for all clauses whose conclusions can be unified with p(a). For example, the clause

$$p(x) \leftarrow q(x) \land r(x)$$

would be selected. In contrast, the following clauses would not be retained:

$$s(x) \leftarrow t(x) \land u(x)$$
$$p(b) \leftarrow v(x)$$

The premises of the selected clauses – also called "bodies" – are in turn similarly processed.

A particular form of backward reasoning, based on linear resolution (see, e.g., [Sti86]), gives the principle – or the so-called operational semantics – of the *Prolog* programming language [Llo87]. Logic programming is however neither restricted to linear resolution, nor to backward reasoning. In order to correctly handle recursive programs, a few new procedures have been defined in the last few years, among others the *magic sets method* [BMSU86, BR87] and SLDAL-resolution [Vie89]. It has been recently observed that these methods all implement the same nonlinear backward reasoning principle [Bry89b, Man89].[1] Forward reasoning is for example applied in abstract interpretation [GCS88, BD88].

It is worth noting that logic programming (without negation) cannot prove all theorems of (negation-free) logical theories. From the clauses

$$p(a)$$
$$p(f(x)) \leftarrow p(x)$$

it is possible to derive the following property:

$$\forall n \in N^* \ p(f^n(x))$$

This universal formulas is however not derived from a logic program, because the induction principle is missing in logic programming.

---

[1]Some methods *implement* backward reasoning in a language of clauses that are processed forward. This property of the *programming language* used does however not characterize the *implementation*.

Disjunctive formulas like a ∨ b cannot be expressed in logic programs. (The logical dependency of a fact a from b or from c can however be expressed by means of two clauses a ← b and a ← c.) In logic programming, a disjunctive statement

$$a \lor b$$

can only be proven by proving a or by proving b. Classical logic gives rise to other proofs, for example based on the *tertio non datur* or excluded middle principle.

The above-mentioned restrictions are all constructive. The formalization of logic programming in full classical logic is already questionable for programs without negation. An interpretation in constructive logic would be more natural.

3. A Non-Classical Inference Principle: Negation as Failure

In logic programming, a negated formula ¬F is considered proven when F is not provable. This inference principle is called *negation as failure* [Cla78]. This manner to derive negative information can be called a reflective inference principle, for the reasoning system reflects on its own knowledge. It departs from classical logic which treats negative and positive expressions similarly.

The negation as failure principle is rather intuitive. It is often used in every-day life. For example, one concludes that there is no direct flight from Nürnberg to Stuttgart if one finds none in the time table. This reasoning is nonmonotonic: If new flights are created, negative conclusions might be no longer valid.

The negation as failure principle has been introduced for relational databases under the name of *closed world assumption* in the seventies [Rei78]. Because it is natural and also because it gives rise to an efficient backward evaluation of negated expressions, this principle has been retained. It is quite interesting to notice that inference principles that are, like negation as failure, unconventional for mathematicians, but rather intuitive for nonmathematicians, have been proposed within constructive logics.

Negation as failure is very close to default reasoning [Rei80]. One can for example see the condition r(x) in the clause p(x) ← q(x) ∧ ¬r(x) as an exception for the negation-free clause p(x) ← q(x). Since negation as failure is a reflective inference principle, it is also related to modal nonmonotonic logics – e.g., autoepistemic logic [Moo85].

Negation as failure poses problems for several reasons. In classical logic, expressions such as a ⇐ b and a ∨ ¬b are equivalent. As a consequence, implications like

$$p \Leftarrow r \land \neg q$$

and

$$q \Leftarrow r \wedge \neg p$$

are also equivalent. Are however the following clauses also equivalent?

$$p \leftarrow r \wedge \neg q$$
$$q \leftarrow r \wedge \neg p$$

Obviously not: From a program consisting of the fact r and first clause, p and $\neg q$ are derived. In contrast $\neg p$ and q follow from r and the second clause.

In constructive logic, the formulas $a \Leftarrow b$ and $a \vee \neg b$ are not equivalent. The implication $a \Leftarrow b$ means only that a can be proven as soon as b is provable. If implications are only interpreted in this constructive manner, an implication $a \Leftarrow \neg a$ is always false: a and $\neg a$ cannot be both provable. The constructive interpretation of implications gives a natural explanation, why the two clauses considered above do not have the same effect: They are not constructively equivalent.

Another difficulty caused by the negation as failure principle is the nonmonotonicity of the operator T. Consider the program $P = \{p \leftarrow \neg q, q \leftarrow r, r\}$. Since $q \notin P$ and $r \in P$, we have $T(P) = \{p, q\}$. The derivation of p is however questionable because it relies on $\neg q$. The later derivation of q contradicts this assumption. The operator T therefore is not suited to defining the denotational semantics of programs with negation. In the next section, we propose a monotonic fixpoint procedure for such logic programs.

4. Monotonic Reasoning on Programs with Negation

In order to restore monotonic reasoning on programs with negation, one has to delay the evaluation of negated expressions. Instead of generating facts, as does the operator T, we propose to collect the negative premises for later evaluations.

The facts a and b and the clause $c \leftarrow a \wedge b \wedge \neg c$ give for example rise to deriving the clause $c \leftarrow \neg c$. From the two following clauses

$$a \leftarrow b \wedge \neg c$$
$$b \leftarrow \neg d$$

one can derive the clause $a \leftarrow \neg c \wedge \neg d$. Nonground clauses are appropriately instanciated and similarly processed.

We first illustrate on an example the operator based on this principle. Then, we recall the formal definition from the article [Bry89a]. Since the new operator generates a kind of conditional facts, we call it "Conditional Fixpoint Operator" and denote it T_c.

Let P be the following logic program:

$$p(x) \leftarrow q(x) \wedge \neg t(x) \wedge \neg r(x) \qquad s(a)$$
$$q(x) \leftarrow s(x) \wedge \neg t(x) \qquad\qquad s(b) \qquad u(b)$$
$$r(x) \leftarrow s(x) \wedge \neg u(x) \qquad\qquad\qquad\qquad u(c)$$

During a first phase, the following clauses – or conditional facts – are generated, until a fixpoint is reached.

$T_c \uparrow^1(P):$ P

     $q(a) \leftarrow \neg t(a)$

     $q(b) \leftarrow \neg t(b)$

     $r(a) \leftarrow \neg u(a)$

     $r(b) \leftarrow \neg u(b)$

$T_c \uparrow^2(P):$ $T_c \uparrow^1(P)$

     $p(a) \leftarrow \neg t(a) \wedge \neg r(a)$

     $p(b) \leftarrow \neg t(b) \wedge \neg r(b)$

Then, in a second phase, these clauses are simplified. The literals $\neg t(a)$ and $\neg t(b)$ are eliminated, since there are neither facts nor clauses defining t. Similarly, $\neg u(a)$ is eliminated. The r(a) and the clause $p(a) \leftarrow \neg r(a)$ result from these first eliminations. Since r(a) has been already obtained, this clause is useless: We eliminate it as well. Pursuing this elimination process results in the following set of facts:

$$\{q(a),\ q(b),\ r(a),\ p(b)\}$$

The definition of the T_c operator given below makes use of a few notations. Given a literal or a conjunction of literals B, 'pos(B)' will denote the conjunction of positive literals in B or, if this is the case, the only positive literal in B. Similarly, 'neg(B)' will denote the conjunction of negative literals or the single negative literal in B. If B does not contain any positive literal (negative literal, resp.) 'pos(B)' ('neg(B)', resp.) is defined as 'true'. A clause, the body of which is 'true' or a conjunction of negative literals, will be called a "conditional fact".

<u>Definition 1:</u>

Let P be a logic program. A conditional fact

$$C:\ A\sigma \leftarrow neg(B\sigma)\ C_1 \wedge ... \wedge C_i \wedge ... \wedge C_n$$

is a conditional consequence of P – i.e., $C \in T_c(P)$ – if:

1. There is a clause $(A \leftarrow B)$ in P
2. There is a substitution σ instanciating the variables in $(A \leftarrow B)$ with ground terms of the language of P
3. $pos(B)\sigma = true$
or $pos(B) = A_1 \wedge ... \wedge A_i \wedge ...A_n$ ($n \geq 1$) and for all $i \in \{1, ..., n\}$ there is a clause $A_i \leftarrow C_i$ in P, or $C_i = true$ and A_i is a fact in P.

Since negative literals are not evaluated during the computation of $T_c(P)$, we have:

Proposition:

The operator $T_c(P)$ is monotonic. It has a unique fixpoint.

The Conditional Fixpoint Procedure is defined as follows:

Definition 2:

Let P be a logic program (without function symbols). The Conditional Fixpoint Procedure consists of two successive phases:
1. The fixpoint $T_c(P)$ is first computed.
2. The set $T_c(P)$ is reduced to a set of facts by applying the following rewriting rules:

$$
\begin{array}{llll}
(F \leftarrow \text{true}) & \rightarrow F & \text{true} \wedge F & \rightarrow F \\
F \wedge \text{true} & \rightarrow F & F \wedge \text{false} & \rightarrow \text{false} \\
\text{false} \wedge F & \rightarrow \text{false} & (F \leftarrow \text{false}) & \rightarrow \Lambda \\
& \neg F & \rightarrow \text{true} &
\end{array}
$$

if F is neither a fact, nor the conclusion of a clause.

The reduction phase of the Conditional Fixpoint Procedure always terminates, as soon as the program under consideration is constructively consistent [Bry89a]. This condition prevents that a clause, the premise of which is the negation of the conclusion – e.g., $p(a) \leftarrow \neg p(a)$ – is derivable. In the next section, we show the procedure of Definition 2 can be adapted to also handle logic programs that are constructively inconsistent.

The extension of Definition 2 to logic programs with function symbols is possible. It requires however more complex notations. For the sake of simplicity, we do not give it here.

The operator T_c does not perform nonmonotonic deductions. However, consequences of a program P can still become invalid in a program extending P with additional facts or clauses.

During the last few years, various classes of syntactically restricted logic programs with negation have been proposed, among others hierarchical, stratified [ABW88, Van88], and locally stratified [Prz88] programs. The constructive formalization of logic programming gives a rather simple motivation for these syntactical restrictions: There are necessary conditions for constructive consistency [Bry89a]. Previously, these syntactical restrictions were only procedurally motivated.

5. Ternary Logic for Unrestricted Programs

The fixpoint semantics of logic programs with negation described in the previous section in terms of the operator T_c does not take into account logic programs that are constructively inconsistent. In practice, it is however often desirable to accept such programs by somehow "ignoring" their inconsistent parts. In this section, we propose a semantics for general programs which formalizes this intention of "overseeing" inconsistencies.

This semantics is defined by slightly modifying the reduction phase of the Conditional Fixpoint Procedure. Clauses such as $p(a) \leftarrow \neg\, p(a)$ that are false from the constructive viewpoint are simply eliminated, and the facts that are "defined" by such clauses – $p(a)$ in case of $p(a) \leftarrow \neg\, p(a)$ – are marked as "unknown". They are thus distinguished from facts without any definition. The negation as failure principle is maintained by interpreting as "false" those facts that have neither consistent, nor inconsistent definitions. We first illustrate this modification of the Conditional Fixpoint Procedure on an example.

Let p be the following logic program:

$$p \leftarrow a \qquad\qquad q \leftarrow \neg p \qquad\qquad a$$
$$p \leftarrow q \qquad\qquad r \leftarrow \neg\, r$$

From P the following marked facts are derived: a: true, p: true, r: unknown, and q: false. This semantics is called ternary or 3-valued, for it has three truth values. The following inequalities express, in the usual formalism of ternary logics, the precedences between the truth values:

$$\text{true} \geq \text{unknown} \geq \text{false}$$

A fact is not assigned the truth value "false" as soon as it can be proven – i.e., marked "true" – or marked "unknown". Similarly, as soon as a fact is proven, it is removed from the facts marked "unknown".

We call "Ternary Fixpoint Procedure" the reasoning informally described above. Like the Conditional Fixpoint Procedure, it is based on the T_c operator. It is formally defined as follows:

Definition 3:
Let P be a logic program (without function symbols). The Ternary Fixpoint Procedure consists of three successive phases:
1. The fixpoint $T_c \uparrow(P)$ is first computed.
2. The set $T_c \uparrow(P)$ is then reduced to a set of marked facts by applying the following rewriting rules, where F denotes an atom:

$$\begin{array}{llll}
F & \rightarrow (\text{F: true}) & (\text{F} \leftarrow \text{true}) & \rightarrow (\text{F: true}) \\
\text{true} \wedge F & \rightarrow F & F \wedge \text{true} & \rightarrow F \\
F \wedge \text{false} & \rightarrow \text{false} & \text{false} \wedge F & \rightarrow \text{false} \\
(\text{F} \leftarrow \text{false}) & \rightarrow \Lambda & & \\
(\text{F} \leftarrow \neg F) & \rightarrow (\text{F: unknown}) & & \\
& \neg F & \rightarrow \text{true} &
\end{array}$$

if F is neither a fact, nor a fact marked "true", nor the conclusion of a clause in P

3. Finally, all facts that are neither marked "true", nor "unknown" are marked "false"; moreover "unknown" marks are removed from facts also marked "true".

It is worth noting that ternary logics are constructive logics. Ternary logics for unrestricted logic programs have been defined in various ways, for example in [Fit85, GL88, VRS88, BF90, Van89]. They do not significantly differ from the above-defined semantics. Moreover, the Ternary Fixpoint Procedure can easily be adapted to reflect these differences. Some authors make use of similar concepts as the T_c operator, which was introduced in [Bry89a].

Like the Conditional Fixpoint Procedure, and in the same way, the Ternary Fixpoint Procedure can be extended to logic programs with function symbols.

6. Application: Reflective Reasoning

Constructive logic is not platonistic. Constructive reasoning does not refer to any meaning of the formulas, but to the formulas themselves. A frequent example of this way of thinking is the hypothesis that distinct constants denote distinct objects. This hypothesis is called *Unique Name Axiom* in database theory. In logic programming, this hypothesis is conveyed through the restriction to Herbrand models. It is questionable, whether human thinking is restricted in a similar way. We shall not try to answer this philosophical issue. We just notice that constructive reasoning is well suited to artificial intelligence: A computer just constructs.

From the constructive viewpoint, it is rather natural to define logic programs with variables ranging over formulas. Such programs are called "meta-programs" [SS86b]. They are often seen as second-order theories. The constructive viewpoint, however, suggests another formalization. Since the variables of meta-programs represent formulas, but no mathematical objects defined by these formulas, meta-programs can be formalized in first-order logic – see in particular [End72], pp. 281-289.

Meta-programming is a mean to overcome restrictions of logic programming, for it gives rise to specifying within logic programming, reasoning principles different from that of logic programming. We illustrate this point by giving some examples of meta-programs that implement some forms of reflective and nonmonotonic reasoning.

Logic programming does not give rise to deriving universal formulas. The following meta-program specify an evaluation of universal formulas in implicative form. The specification is sound and complete for restricted universal quantification [Bry89c].

$$\text{forall}(x, y \Rightarrow z) \quad \leftarrow \quad \neg (y \wedge \neg z)$$

Let F be the formula $\forall x \; p(x) \Rightarrow q(x)$ and P the program $\{p(a), p(b), q(a), q(b), q(c)\}$. When y is instanciated with $p(x)$, and z with $q(x)$, a backward evaluation of forall(x, y $\Rightarrow$ z) binds the variable x successively with all values in p-facts. For each instantiation of x, $q(x)$ is checked. It is worth noting, that this way to prove universal formulas is constructive.

We show by means of a simple example how a form of autoepistemic reasoning can be easily implemented by a logic meta-program.

Let A, B, and C be three agents. assume that A believes all what B believes, provided that C does not believe it. Assume also that B believes everything provable. Assume finally that C believes everything simple. We assume that an information is simple if it is provable without applying *modus ponens*.

$$\text{believes}(A, x) \leftarrow \text{believes}(B, x) \wedge \neg \text{believes}(C, x)$$
$$\text{believes}(C, x) \leftarrow \text{fact}(x)$$
$$\text{believes}(B, x) \leftarrow \text{proven}(x)$$
$$\text{believes}(C, x) \leftarrow \text{simple}(x)$$

$$\text{simple}(F_1 \wedge F_2) \leftarrow \text{simple}(F_1) \wedge \text{simple}(F_2)$$
$$\text{simple}(F) \qquad \leftarrow \text{fact}(F)$$

Finally, we propose a meta-program for default reasoning. General rules are assumed to be stored is facts by means of a predicate "clause". For example,

$$\text{clause}(\text{flies}(x) \leftarrow \text{bird}(x))$$

means that birds normally fly. Exceptions to general rules are assumed to be expressed by using an "except" predicate. Since penguins are birds, but do not fly, we have:

$$\text{except}(\text{flies}(\text{penguin}))$$

The following meta-program implements default reasoning:

$$\text{proved}(x) \qquad \leftarrow \text{clause}(x \leftarrow y) \wedge \text{proved}(y) \wedge \neg \text{except}(x)$$
$$\text{proved}(x_1 \wedge x_1) \leftarrow \text{proved}(x_1) \wedge \text{proved}(x_2)$$
$$\text{proved}(x) \qquad \leftarrow \text{fact}(x)$$

From the following object-program

$$\begin{aligned}
&\text{clause(flies(x)} \leftarrow \text{bird(x))} \\
&\text{bird(crow)} \\
&\text{bird(penguin)} \\
&\text{except(flies(penguin))}
\end{aligned}$$

the above-defined meta-program derives only the fact flies(crow).

It is worth noting that exceptions – i.e., "except" facts – can as well be defined by means of clauses. These clauses in turn can admit exceptions. Processing such "nested exceptions" by means of the Conditional or Ternary Fixpoint Procedures results in the hierarchical reasoning suggested by Poole in [Poo98] for default reasoning systems. Thus, logic meta-programming is very convenient to easily specifying – and implementing – sophisticated default logics.

We illustrate the notion of "nested exceptions" on an example. Cars must stop if the traffic light is red. This rule however does not apply to ambulance cars provided they are in service. This last restriction – i.e., "in service" – may be viewed as an exception to the derogatory rule for ambulance cars. In the formalism of the above-defined meta-program, this example can be represented as follows:

$$\begin{aligned}
\text{stop} &\quad \leftarrow \text{red} \\
\text{except(stop)} &\quad \leftarrow \text{ambulance} \\
\text{except(except(stop))} &\quad \leftarrow \text{after-service}
\end{aligned}$$

Instead of refering to predicates – "stop" and "except(stop)" in the previous example – exceptions can also refer to clauses. This requires a slight modification of the formalism. This is easily achieved by, for example, naming the clauses and modifying the "proved" meta-program. We assume that a clause $x \leftarrow y$ is uniquely assigned an identifier z. We also assume that clauses named in this way can be retrieved using a binary predicate "clause":

$$\text{clause}(z, x \leftarrow y)$$

The following meta-program handles exceptions to clauses:

$$\begin{aligned}
\text{proved}(x) &\quad \leftarrow \text{clause}(z, x \leftarrow y) \land \text{proved}(y) \land \neg \text{except}(z) \\
\text{proved}(x_1 \land x_1) &\quad \leftarrow \text{proved}(x_1) \land \text{proved}(x_2) \\
\text{proved}(x) &\quad \leftarrow \text{fact}(x)
\end{aligned}$$

Doing so, different exceptions can be given for "stop", depending on the stop-rule under consideration. Such a refined representation give rises, for example, to express that there is no derogation to stopping at level crossing when a train arrives. The following clauses formalize this example:

```
clause(1,       stop          ← red)
clause(2,       stop          ← level-crossing ∧ train)
clause(3, except(1)           ← ambulance)
clause(4, except(except(1))) ← after-service)
```

Clearly, some applications might require to handle both, exceptions to clauses as well as exceptions to predicates, depending on the predicates. We leave specifying the corresponding meta-program to the reader, for it is quite easy.

Because it gives rise to easily refining various deduction schemes, we argue that meta-programming is very convenient for defining and implementing complex reasoning, especially nonmonotonic reasoning systems.

7. Conclusion

In this article, we have proposed a particular interpretation of logic programs. According to this interpretation, logic programming departs from modern mathematics. It restricts reasoning to certain inference rules, in a manner which corresponds to a primitive form of mathematical logic, constructive logic. It provides us with rather natural explanations for several features of logic programming.

In particular, it explains very naturally syntactical restrictions such as "stratification" and "local stratification" that were proposed for overcoming difficulties resulting from the non-monotonicity of negation as failure. The constructive interpretation of logic programs also provides us with a convenient framework for defining a ternary semantics for syntactically unrestricted programs.

The constructive viewpoint also suggests a formalization of meta-programming. By means of a few examples of meta-programs, we have shown how logic programming is closely related to other non-monotonic logics. We have shown how meta-programming gives rise to elegant specifications of various logics that depart from the very semantics of logic programming. These specifications are as well implementations. Thus, although it implements a restricted, constructive fragment of classical logic, logic programming gives rise, through meta-programming, to overcoming its own limitation.

We argue that it is preferable to specify refined forms of reasoning as logic meta-programs, instead of developing new formalisms or new implementation paradigms, for the following reasons. First, constructive reasoning – the paradigm of logic programming – is very natural and intuitive. Second, the meta-programming approach gives rise to easily combining into one system, different forms of reasoning. Finally, program transformation techniques – in particular partial evaluation, see, e.g., [SS86a] – have been developed for automatically generating efficient implementations from specifications in form of logic meta-programs.

Acknowledgement

This research was partly supported by the European Community in the framework of the Basic Research Action *Compulog* No. 3012.

References

[ABW88] K. R. Apt, H. A. Blair, and A. Walker. *Foundations of Deductive Databases and Logic Programming*, chapter Towards a Theory of Declarative Knowledge. Morgan Kaufmann, Los Altos, Calif., 1988.

[BD88] M. Bruynooghe and D. De Schreye. Tutorial Notes for: Abstract Interpretation in Logic Programming. Research report, University of Leuven, 1988. Invited tutorial of the 5th Int. Conf. on Logic Programming (ICLP), Seattle, 1988.

[Bec88] C. Beckstein. *Zur Logik der Logik-Programmierung – Ein konstruktiver Ansatz*. Informatik-Fachbericht 199. Springer-Verlag, 1988.

[BF90] N. Bidoit and C. Froidevaux. Negation by Default and Unstratifiable Logic Programs. *Theoretical Computer Science*, 1990. To appear.

[BMSU86] F. Bancilhon, D. Maier, Y. Sagiv, and J. Ullamn. Magic Sets and other Strange Ways to Implement Logic Programs. In *Proc. 5th ACM SIGMOD-SIGACT Symp. on Principles of Database Systems (PODS)*, 1986.

[Boj86] D. Bojadziev. A Constructive View of Prolog. *Journal of Logic Programming*, 3(1):69–74, 1986.

[BR87] C. Beeri and R. Ramakrishnan. On the Power of Magic. In *Proc. 6th ACM SIGMOD-SIGACT Symp. on Principles of Database Systems (PODS)*, 1987.

[Bry89a] F. Bry. Logic Programming as Constructivism: A Formalization and its Application to Databases. In *Proc. 8th ACM SIGACT-SIGMOD-SIGART Symp. on Principles of Database Systems (PODS)*, Philadelphia, Penn., March 1989.

[Bry89b] F. Bry. Query Evaluation in Recursive Databases: Bottom-up and Top-down Reconciled. In *Proc. 1st Int. Conf. on Deductive and Object-Oriented Databases (DOOD)*, Kyoto, Japan, Dec. 1989. The complete version of this article will appear in the special issue of *Data & Knowledge Engineering* devoted to the DOOD '89 Conference.

[Bry89c] F. Bry. Towards an Efficient Evaluation of General Queries: Quantifier and Disjunction Processing Revisited. In *Proc. ACM-SIGMOD Conf. on Management of Data (SIGMOD)*, Portland, Oreg., May-June 1989.

[Bry90]   F. Bry. Negation in logischer Programmierung: Eine Formalisierung in konstruktiver Logik. Research Report IR-KB-72, ECRC, 1990.

[Cal79]   A. Calder. Constructive Mathematics. *Scientific American*, 241(4):134–143, 1979.

[Cla78]   K. L. Clark. *Logic and Databases*, chapter Negation as Failure. Plenum Press, New York, 1978.

[dK86]   J. de Kleer. An Assumption-Based Truth Maintenance system. *Artificial Intelligence*, 28:127–162, 1986.

[Doy79]   J. Doyle. A Truth Maintenance system. *Artificial Intelligence*, 24, 1979.

[End72]   H. B. Enderton. *A Mathematical Introduction to Logic*. Academic Press, 1972.

[Fit85]   M. Fitting. A Kripke-Kleene Semantics for Logic Programs. *Journal of Logic Programming*, 4:295–312, 1985.

[GCS88]   J. Gallagher, M. Codish, and E. Shapiro. Specialisation of Prolog and FCP Programs Using Abstract Interpretation. *New Generation Computing*, 6:159–186, 1988.

[Gir87]   J.-Y. Girard. Linear Logic. *Theoretical Computer Science*, 50:1–102, 1987.

[GL88]   M. Gelfond and V. Lifschitz. The Stable Model Semantics for Logic Programming. In *Proc. 5th Int. Conf. and Symp. on Logic Programming (ICLP-SLP)*, Seattle, 1988.

[GLT89]   J.-Y. Girard, Y. Lafont, and P. Taylor. *Proofs and Types*. Cambridge Tracts in Theoretical Computer Science 7. Cambridge University Press, 1989.

[Hue86]   G. Huet. *Fundamentals of Artificial Intelligence*, chapter Deduction and Computation. LNCS 232. Springer-Verlag, 1986.

[Kow79]   R. Kowalski. *Logic for Problem Solving*. North Holland, 1979.

[Llo87]   J. Lloyd. *Foundations of Logic Programming*. Symbolic Computation. Springer-Verlag, 1987. Second Edition.

[Man89]   R. Manthey. Can We Reach a Uniform Paradigm for Deductive Query Evaluation? In *Proc. 3rd Int. GI Congress "Knowledge Based Systems"*, München, Oct. 1989.

[Moo85]   R. C. Moore. Semantical Considerations on Nonmonotonic Logic. *Artificial Intelligence*, 25, 1985.

[Poo98]   D. Poole. Explanation and Prediction: An Architecture for Default and Abductive Reasoning. Tech. Report 89-4, Dept. of Comp. Sc., Univ. of British Columbia, 198.

[Prz88]   T. C. Przymusinski. *Foundations of Deductive Databases and Logic Programming*, chapter On the Declarative Semantics of Deductive Databases and Logic Programs. Morgan Kaufmann, Los Altos, Calif., 1988.

[Rei78]   R. Reiter. *On Closed World Databases*, pages 56–76. Plenum Press, New York, 1978.

[Rei80]   R. Reiter. A Logic for Default Reasoning. *Artificial Intelligence*, 13, 1980.

[Rob65a] J. A. Robinson. A Machine-Oriented Logik Based on the Resolution Principle. *Journal of the ACM*, 12(1):23–41, 1965.

[Rob65b] J. A. Robinson. Automated Deduction with Hyper-Resolution. *Journal of Comp. Math.*, 1:227–234, 1965.

[SS86a]   S. Safra and E. Shapiro. Meta Interpreters for Real. In *Proc. 12th IFIP World Congress*, 1986.

[SS86b]   L. Sterling and E. Shapiro. *The Art of Prolog*. MIT Press, 1986.

[Sti86]   M. E. Stickel. *An Introduction to Automated Deduction*, pages 75–132. Springer Verlag, Berlin, 1986.

[Van88]   A. Van Gelder. *Foundations of Deductive Databases and Logic Programming*, chapter Negation as Failure Using Tight Derivations for General Logic Programs. Morgan Kaufmann, Los Altos, Calif., 1988.

[Van89]   A. Van Gelder. The Alternating Fixpoint of Logic Programs with Negation. In *Proc. 8th ACM SIGACT-SIGMOD-SIGART Symp. on Principles of Database Systems (PODS)*, Philadelphia, Penn., March 1989.

[vEK76]   M. van Emden and R. Kowalski. The Semantics of Predicate Logic as a Programming Language. *Journal of the ACM*, 23(4):733–742, 1976.

[Vie89]   L. Vieille. Recursive Query Processing: The Power of Logic. *Theoretical Computer Science*, 69(1):1–53, Dec. 1989.

[VRS88]   A. Van Gelder, K. A. Ross, and Schlipf J. S. The Well-Founded Semantics for General Logic Programs. In *Proc. 7th ACM SIGACT-SIGMOD-SIGART Symp. on Principles of Database Systems (PODS)*, Austin, Texas, March 1988.

Database Support for Knowledge Representation?

G. Görz, University of Hamburg, Computer Science Department

SO — you may well ask — WHAT? Here's what: the development of large, distributed AI systems, subject to access by many different people and systems, requires that system developers look to the Database community for help and guidance. It would be idiotic for people in AI to reinvent all those wheels. Moreover, as reinvented, they'd probably end up as irregular polyhedra. On the other hand, I think it's a mistake to think of Databases *transmuting themselves* — growing up — *into something they're not. If one conceives the* domain closure *and* closed world assumptions, *or of the subsumption relations embodied in* IS-A *hierarchies as being explicitly expressed, say by way of first order sentences, then there is no longer thinking about* Databases... Relational Databases ... *cannot gain the world without thereby losing their soul.*

David Israel [20]

1 Introduction: Semantic Processes

As the complexity of applications of knowledge-based systems grows, there can be no doubt that these systems deserve support by Database Management Systems technology. But there is yet no definite answer on how both technologies can be brought towards a fruitful interaction. This paper is intended as a contribution to the discussion and will as such take up a few — at least in the "knowledge representation community" — quite common arguments to make a more precise suggestion what a useful interface between both might be and how it could be organized. Interestingly, there seem to be similar considerations in the "database management community": I see a considerable correspondence with my view in e.g. the contributions of Mattos et al. [1]

The biggest deficiency of the current discussion seems to be its limitation by the use of different terminologies and different views of the technology. Therefore, I will try to sketch the premises behind the arguments in this paper in its first two sections.

Our basic assumption is that the technology of symbolic information processing can provide help to the solution of problems which come up in modern societies. In this sense, the tradition of formal reasoning is taken up by this technology in a constructive way and implemented on "mechanical" devices. Of course, we do not claim that *symbolic information processing* is the only way, but a fruitful approach. In our understanding, symbolic information processing is built upon a reconstruction of knowledge and reasoning in symbolic form. As soon as we speak of knowledge — even if we ascribe "implicit knowledge" — we automatically assume that it can be spelled out in a propositional form.

This view has been clearly worked out by Brian Smith [41] who states that computer science should be considered as the study of *semantic processes* and not, fundamentally,

---

[1]this volume, cf. [13, 32]

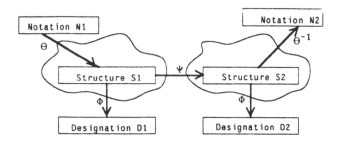

Figure 1: A framework for computational semantics after B. Smith

a study of language, of something static. Computers, programming languages, and programs service the description, construction, and interaction with semantic processes of various sorts. Computer science primarily studies processes, and secondarily the languages used to describe and interact with them. Processes are called "semantic", because they have content, they are about something — in other words, they are *significant*. Symbolic systems of information processes are created to represent and transport meaning, to be *interpreted*, as opposed to other technical systems and devices (like toasters) which we don't interpret. So the question for the semantics of symbolic systems is central to computer science. In particular, we have to consider the encoding of meaning in the structure of formal systems as well as extensional and intensional properties of processes and languages.

Basically, we have to consider the mapping between names and expressions to their denotations and vice versa as well as the realization of logical decisions by means of formal devices. To do that, we have to distinguish between

1. (real and abstract) objects and events, in which a computational process is embedded;

2. internal elements, structures, and processes in a computational system which are formal objects; and

3. expressions which serve the communication between the environment — the external world — and the internal world of a computational system.

In Fig. 1, the level in the middle represents the level of processes, the syntactic domain of program and data structures, the "structural field", which are interpreted by a processor. The "internalization" mapping θ maps external expressions on internal structures of a computational system, the mapping ψ realizes the transformation between different internal states of the system, its "behavior". The interpretation mapping ϕ relates internal structures to their designations in the external world, i.e. it provides the link between the syntactic and the semantic level. The interdependence of ϕ and ψ deserves particular attention — in logical systems, it is just the equivalence of proof theoretic consequence ($\vdash$) and semantic entailment ($\models$).

In the following I will give an overview of some fundamental issues of knowledge representation and then try to remind of a view on Database Management Systems from a specific standpoint within the field of knowledge representation which is best characterized by the well-known term "knowledge level"[2]. Under this view, the interaction between knowledge representation and databases is much more fundamental than just that each can contribute algorithms and techniques to the other. Instead, databases are interpreted as knowledge bases of a certain limited form. As Brachman and Levesque argue, this limitation in representation form can be motivated by a fundamental tradeoff that all knowledge representation and reasoning systems are faced with.

The question of integration, then, will be tackled from the perspective of the integration of how different programming styles. In any case, we suggest that the database component should play the role of a server within complex problem solving systems. Basically, it will depend on the available tools where the border line between the — programming and representation — language system and the database system will be drawn.

2 Knowledge Representation

Knowledge Representation in its full generality is best characterized by what Smith [41] has called the *Knowledge Representation Hypothesis:*

'Any mechanically embodied intelligent process will be comprised of structural ingredients that

- we as external observers naturally take to represent a propositional account of the knowledge that the overall process exhibits, and

- independent of such external semantical attribution, play a formal but causal and essential role in engendering the behavior that manifests that knowledge."

Following this hypothesis, there are two major properties which a knowledge representation system have to satisfy. To quote Brachman and Levesque [10], it must be possible to interpret its structures *propositionally*, i.e. as expressions of a language which has a *truth theory*. Secondly, these structures must necessarily play a *causal role* in the behavior of that system — which should agree with our understanding of them as propositions representing knowledge.

A *knowledge representation formalism* is a formal language with a well defined syntax and semantics, a *knowledge representation system* is an implementation of such a formalism which supports

- the *interpretation* of well formed expressions of a knowledge representation formalism by sound and — in the ideal case — complete *inference algorithms*, and

- the *administration* and *actualization* of formally represented knowledge bases.

In addition, such a system should provide some kind of expressive adequacy, i.e. it should be able to express all semantically required distinctions, and the notation should be efficient. Now, one might argue that first order logic is just such a formalism: Its symbolic structures are the well formed logical expressions and its operations can be defined in a purely formal way — specified by its proof theory — and they preserve semantic properties — as specified by its proof theory. The equivalence of proof theoretic consequence

---

[2]cf. Brachman and Levesque [10]

(derivability ⊢) and semantic entailment (logical consequence ⊨) guarantees that the fundamental requirements are fulfilled. So, where is the problem? The problem is, that first order logic is undecidable. Even if the decision problem is made solvable by restricting the formalism, it can in general not be solved in realistic time (co-NP-hardness).

So we have to accept that, if we still consider knowledge representation as a project worthwhile to pursue, we have to look for a compromise (cf. Levesque [27, 29]). In fact, three possibilities have been suggested in order to reduce the general computational needs:

- *Special languages* by reducing the expressive power, in particular w.r.t. the representability of incomplete knowledge. Examples of this are relational databases, logic programs, semantic networks and frame languages.

- *Limited inference* which is a form of implication weaker than full logical consequence. This means that additional premises on incomplete knowledge bases have to be introduced, e.g. worlds, situations, beliefs, etc. which can be realized by some version of modal logic.

- *"Defaults" and assumptions:* Incomplete knowledge bases are completed by assumptions made on the basis of absence of contradicting evidence which results in the introduction of non-monotonic inference.

For the remainder of this section, we will concentrate on the first possibility by considering *object-centered representations*. This type of representation can be characterized by the fact that knowledge bases are organized in a way that either primary objects or concepts ("what is"), or actions, including inference relations between objects ("how to") are focussed. Historically, semantic networks or frame languages are typical representatives of this approach. We will illustrate it by a prototypical *attributive representation system* which is a sort of kernel language for these approaches and can, on the other hand, be seen as a notationally efficient syntax for a subset of first order logic terms. Its peculiarity consists in the treatment of unary and binary predicates — predicates of a higher arity are not admitted: Unary predicates are called *types,* binary predicates *attributes.* Types are ordered along a *taxonomic hierarchy,* which defines an inheritance relation. It is possible to formulate restrictions on attributes for instances of a type. An essential achievement of this type of representation is that a knowledge base can be represented as a directed labeled graph whose nodes denote constants or types and whose arcs denote attributes or IS-A relations. Then, certain inferences, namely the taxonomic ones, can be realized efficiently as graph search algorithms!

The basic expressions in such an object centered representation system are called *attribute terms.* They describe or classify objects of a given domain of discourse by sets of attributes which are interpreted as partial functions (from attribute names to their values which may either be atomic or attribute terms themselves). Identity of values can be expressed by *coreference* tags. Viewing attribute terms as specifications of complex types, a *lattice* of such types can be built up by means of a *subsumption* relation which is ordered either by information content or, inversely, by the set of models they describe. Compatible attribute terms, viewed as partial descriptions of objects of the domain of discourse, can be combined to more specific descriptions by an operation known as *graph unification.*

We will illustrate that with an example from computational linguistics, where attributive systems have become very popular within the last ten years. They constitute the basis of *constraint-based* grammar formalisms — also called unification-based grammars

$$\text{present3rdsg} \doteq \left[\begin{array}{l} \text{tense: present} \\ \text{subj:} \left[\begin{array}{l} \text{num: sg} \\ \text{person: 3rd} \end{array} \right] \end{array} \right] \qquad V \longrightarrow \text{sings}$$

$$V: \left[\begin{array}{l} \text{pred: verb: sing} \\ \text{transitive} \\ \text{present3rdsg} \end{array} \right]$$

$$\text{transitive} \doteq \left[\begin{array}{l} \text{agent pred} \downarrow \text{subj} \\ \text{what pred} \downarrow \text{obj} \end{array} \right]$$

Figure 2: A lexical entry using templates

— have been proven to be a valuable tool of linguistic description. In most of these formalisms, grammars are defined in the form of context-free rules annotated by constraint equations which formulate conditions for successful rule application. Descriptions of linguistic objects like phrases, sentences, etc., are obtained as solutions of the system of constraint equations which is built up while the parser applies the grammar rules. Descriptions are often written in a matrix notation which arranges the set of attributes vertically with one attribute (name-value pair) per line. The basic building blocks of these descriptions are taken out of the lexicon; fig. 2 shows an example lexicon entry for the verb "sings"[3].

For the representation of lexical information the representation system should provide an abstraction mechanism to express common properties of classes of words on a general level[4]. To make this requirement clear, consider that e.g., all transitive verbs have the same subcategorization scheme. The attribute term notation can easily be extended by *templates,* named attribute terms which enable a simple macro-like mechanism to express *simple inheritance* — cf. fig. 2[5]. An inheritance graph can easily be built up through hierarchical definitions such that even multiple inheritance is expressible.

It has been mentioned above that such an attributive description system could also be seen as a sort of kernel language for object centered terminological representation systems of the KL-ONE type[6] which grew out of research in semantic networks and frame systems. In fact, as Nebel and Smolka ([34], p. 1) point out, both approaches — terminological representation systems and constraint grammar formalisms —

1. rely on attributes as the primary notational primitive for representing knowledge

2. are best formalized as first-order logics with Tarski-style models

3. employ compositional set descriptions.

But terminological representation systems differ significantly in

---

[3]example taken from Nebel and Smolka [34]

[4]for a related approach in the field of lexical semantics cf. e.g. Sowa [42]

[5]The lines with down arrows in attribute terms are called agreements which are a convenient way for expressing *coreference* in that they enforce that the strings of attribute names on both sides of the down arrow (called paths) lead to the same node.

[6]cf. e.g. Nebel [35], Schefe [38], Brachman et al. [8], Levesque [27]

1. the representational constructs they provide: agreements generalize to roles and "role value maps" which result in an undecidable subsumption relation on concept descriptions, as well as in their

2. inference operations: set inclusion and set membership relations are inferred on user-defined symbols.

To apply large amounts of knowledge, so called "hybrid representation systems" like KRYPTON, KL-TWO, KANDOR, MESA were developed were one component is a terminological representation system, the "TBox". The other component, the assertional or "ABox", serves for state or situation descriptions, for which either a subset of first order logic or object centered schemata have been used. The prevailing dilemma has been strikingly summarized by Levesque [28]:

- "certain forms of knowledge are inherently intractable and cannot be fully applied within reasonable resource bounds,

- a special kind of knowledge (namely which is complete in content and vivid[7] in form) can be fully applied,

- the application of knowledge can also be made computationally tractable by making it logically unsound and incomplete in a principled way."

3 A "Second Turn" to Knowledge Processing

When the field of "Artificial Intelligence" was initiated in the mid-fifties, two research trends stood at the very beginning: the task to formalize common sense knowledge and the project to develop powerful general problem solving methods. Within the following fifteen years research concentrated mainly on the second task which led in fact to considerable progress in the development of algorithms, in particular for heuristic search. But, generally spoken, this research direction could not gain the expected success, because those general methods did not apply *domain specific* knowledge and heuristics to a satisfactory extent. So, around 1970 a turn to "knowledge processing" was initiated. Instead of emphasizing general methods the application of large amounts of domain specific knowledge and of specific methods were expected to be key for a qualitative leap. Knowledge-based systems which incorporated lots of domain specific knowledge from e.g. engineering, medicine, etc. were constructed to solve diagnostic or configuration tasks. The most common technique to represent and apply knowledge was by means of condition-action rules. Specialized languages for knowledge representation came up like frame or semantic network languages which were suitable to represent large amounts of domain specific knowledge and heuristics in a structured way. The type of modelling which was characteristic for these system is often called "shallow" because the kind of knowledge employed in them consists of large collections of empirical associations. What is still missing was a formalization of common sense knowledge as a basis for expert reasoning. Many "expert systems" are good at standard cases, but they lack what makes truly an expert: imagination, creativity, etc.

Another fifteen year later, around the mid 1980s, one could observe another turn to knowledge processing in the following sense: On the one hand, considerable progress was made in "deep modeling" as well as in the formalization of some kinds of common sense

---

[7]i.e. which can be directly mapped into corresponding structures of the representation system

reasoning, and on the other hand bigger and bigger chunks of empirical knowledge are subject to a formal machine-based representation. One of the common hopes of AI is that once there is enough knowledge and a large number of "agents" specialized to particular problem solving tasks, their interaction can lead to "emergent behavior" which exhibits intelligence of a new quality, intelligence which has not been coded explicitly in its components. Deep modeling techniques allow to describe systems not only by superficial empirical associations, but by causal models from which predictions or explanations on the behavior of a modeled system can be derived by means of quantitative or qualitative simulation. Techniques of modeling spatial and temporal relations, as well as qualitative descriptions of physical processes — which are still a very active research subject — provide the fundamentals for deep modeling. Cyc, an impressive project to provide huge amounts of encyclopedic knowledge systematically by means of modern knowledge representation techniques has been initiated by Lenat and others[8]. Similar developments are undertaken in the field of computational linguistics where a shift from machine-readable dictionaries to the representation of lexical information is happening. The representational capabilities required include a variety of services like classification, forward- and backward-chained inferences, semantic unification, inconsistency detection, an object-oriented programming interface, and the integration of highly tuned and fast special-purpose reasoners, all of which are current research topics in knowledge representation systems.

Nevertheless, it should be recognized that encyclopedic knowledge as well as domain knowledge in the above mentioned sense are mostly of a different type than common sense knowledge which includes intentionality, imagination, intuition, body experience, etc. This remark is not intended to question the whole enterprise as such but instead to relativize too far going hopes and expectations. In particular, the formal representation of encyclopedic knowledge as done in the Cyc project of Lenat, Guha and others [26] could be a basis of yet still unforeseeable usefulness for building a new generation of knowledge-based systems.

As the size of knowledge corpora grows, the use of database technology becomes inevitable. Whereas the sizes of the knowledge bases of most of the earlier knowledge-based systems did fit into the virtual address spaces of average workstation computers, this will definitely no longer hold of truly large systems like Cyc. Knowledge representation systems like those of the hybrid type (cf. last section) provide a lot of representational devices current database management systems don't have, but there is a similar argument in the reverse direction, e.g. with respect to secondary memory management, recovery, etc. (cf. next section). Therefore, from the perspective of knowledge representation, support by database management technology is overdue, and an integration of both technologies is one of the indispensable prerequisites for future knowledge-based systems. A combination with parallel computer architectures[9] will provide a new dimension of the exploitation of these knowledge bases ("memory based reasoning", cf. Stanfill [43]).

4   How Can Database Technology Help?

Since the goal of this paper is to discuss the integration aspect from the knowledge representation perspective, the advancements of database management systems (DBMS)

---

[8]cf. Lenat et al. [23, 24, 25, 26]; see also Weyer [45, 46]
[9]cf. Hillis [18, 19]

technology which are of biggest interest from this view shall only briefly be summarized.[10]
These features include:

- persistency of objects,

- data integration and data independence,

- integrity control,

- secondary memory management, in particular the management of large data volumes,

- interactive query interface and query optimization,

- concurrency management,

- recovery,

- data security,

- mechanisms for data protection, etc.

Unfortunately, the task of integration is much more difficult than just adding to one technology salient features of the other one. Our understanding of integration is not primarily to combine already existing software components and data, but a conceptual combination on the level of functionality. So, if we speak of integrating DBMS features into knowledge representation systems, we first of all have to specify precisely what we mean by *integration*. This will be done in the next section by borrowing a distinction of various integration types from the discussion on the integration of different programming styles.

Before doing that, we have to consider that both technologies grew out of different needs and were shaped according to different requirements. In typical cases, the representations of objects stored in DBMS have a simple static structure, whereas most knowledge representation system applications deal with complex objects, often with a priori underdetermined dynamic structures as well as with complex static or temporal relations between objects — like multiple inheritance, to name just one. Many knowledge representation systems provide context mechanisms and multiple views on representations. On the other hand, the features of data/knowledge persistency and shareability of mass data or concurrent access to knowledge bases, respectively, are becoming necessary for large applications. This includes access of knowledge processing systems to existing databases. So, it seems obvious to view DBMS as a kind of instance servers, where the accessing knowledge representation system transforms database entries into a nonpersistent internal form. Furthermore, we have to consider a difference with respect to the notion of a query. Whereas in a normal DBMS, we have few queries and large answers, in knowledge representation systems typically there are many queries with small answers, such that there may be different requirements for query optimization.

---

[10]as taken from current data and knowledge base management systems literature like Schmidt and Thanos [39] and in particular the paper by Brodie [12] in that volume; cf. also the contributions by Mattos and Walther in this volume

Among the various types of DBMS, object-oriented database systems[11] are the ones which meet the requirements of knowledge representation systems best from a conceptual point of view. They provide similar concepts for the representation of structure and behavior by data objects, for object identity ("surrogates"), object versions, data encapsulation, and inheritance. In contrast to more conventional DBMS approaches like the relational one they allow to keep information about semantic constraints — which in conventional DBMS are often left to the responsibility of the application. Object-oriented database systems support the integration of semantic data models with programming language concepts on the linguistic level, because the underlying object-oriented programming language constructs provide appropriate abstraction mechanisms for knowledge modeling.

5 An Approach to Integration

We will start the discussion of integration from a linguistic viewpoint, i.e. from the programming system perspective. For some years, there is a trend towards the integration of different programming styles. The problem of integration can clearly be illustrated by considering two typical programming styles, the functional and logic-oriented one. Focussing on these two styles has sufficient generality, since the functional programming style can easily be augmented by the relevant features of imperative and object-oriented programming[12]. The *functional programming style* has its origins in the mathematical concept of functions, which is operationalized in functional languages. The basic mode of computation is the evaluation (or normalization and reduction) mode, the application of functions to arguments. The information flow in functional systems is directed, deterministic, and convergent. The *logic-oriented programming style* has its origins in the concept of logical deduction. Its basic mode of computation is deductive[13]. Logic-oriented systems are basically relational; the information flow is multi-directional, non-deterministic, and not necessarily convergent. Functions expect an input and yield an output, but relations don't. Functions don't fail and "backtrack", but relations do. Functions must terminate for all admissible inputs, but relations don't.

From a formal point of view, functional programming is subsumed by relational programming — so what is the meaning of integration? There are good reasons to keep both the evaluation and the deduction modes separate, but on the other hand there are many advantages to have both accessible in one programming system. The deduction mode requires a high processing load, whereas the evaluation mode can be implemented very efficiently. In addition, the evaluation mode allows processing of objects of higher order (higher order functions). But in functional programming, missing arguments cannot be synthesized by unification as in relational programming. To offer both programming styles in one system is also supported by the argument that, depending on the problem type, either functional or relational programming may provide the better adequacy of notation. With respect to efficiency, it must be stated that the control problem in deduction mode — in contrast to the evaluation mode — is yet unsolved. Furthermore, the introduction of the concept of the "logical variable" into functional programming leads to a great gain in expressiveness: syntactically it allows to use parameter expressions instead of simple

---

[11]cf. Dittrich [14], Zaniolo [47]; for a commercially available system see e.g. Symbolics' STATICE [44]
[12]as has been demonstrated for the LISP dialect SCHEME.
[13]cf. the programming language PROLOG

parameter lists, semantically it allows to determine values by intersection of constraints, and operationally "call by unification" becomes possible, which is inherently bidirectional. To summarize, logical variables allow to express object uniqueness by shared access on a high abstraction level.

For the integration of programming styles, the following approaches can be distinguished[14]. Given two programming languages A and B, we have the

1. Embedding approach: Implementation of A in B

   (a) Functional embedding: implementation of A in the system environment of B with mutual function call.

   (b) Environment embedding: A and B have a common binding environment such that identifier references of the embedded language A can directly refer to bindings in B.

   (c) Complete embedding: environment embedding plus control embedding, i.e. at any time the control context is explicitly available. This is our favorite approach.

2. Syntactic approach: The functionality of A is realized by syntactic transformations in B. This approach has some similarity with environment embedding; there are problems with efficiency and the loss of transparency.

3. Algebraic approach: The operational semantics for functional programming is given by term substitution. Problems here are efficiency and the simplification order over terms.

4. Embedding in a higher order logic: mapping into higher order lambda calculus as the common basis of combinatory logic and functional programming — which results at least in a considerable overhead.

An integrated language design on the basis of the complete embedding approach has been introduced by Ait-Kaci et al. under the name LIFE[15]. Besides their implementation in PROLOG there is also one in SCHEME by Backofen and Euler[16] which includes distributive disjunctions in attribute terms. LIFE has three components: an (object-oriented) type component introducing attribute terms, a functional and a relational component. It integrates three sublanguages: A functional sublanguage which combines types and functions is equivalent to SCHEME with attribute terms instead of constructors. A logical sublanguage combines pure PROLOG with attribute terms as generalized records such that type hierarchies can be described by attributive expressions and type inferences can directly be conducted through inheritance relations (subsumption in the type lattice). The third sublanguage combines the relational and functional styles by generalizing logical terms with the introduction of applicative expressions, i.e. interpreted functional terms with logical variables. So delayed unification is poossible until operands are available, i.e. until variable instantiations allow a reduction of applicative expressions. This is captured by the concept of *residuation,* through which such unifications are viewed as residuum equations which are to be proved instead of being solved.

---

[14]cf. Ait-Kaci, Nasr and Lincoln [1, 2, 3, 4], Haynes [17]

[15]cf. Ait-Kaci, Nasr and Lincoln [1, 2, 3, 4]

[16]cf. Backofen, Euler, and Görz [7, 6, 15]

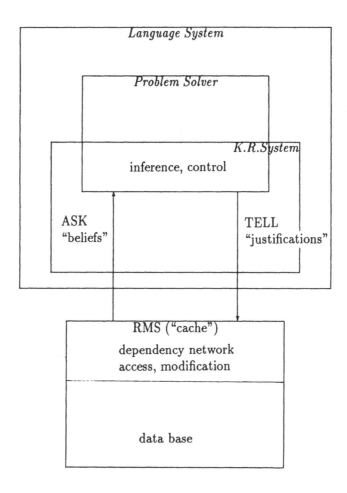

Figure 3: An integrated system architecture

The final thesis of this paper is that such a language design is immediately suitable for a close integration of databases on the functional level. Under this view, a database is characterized *functionally* — what it can be asked (ASK operator) and what it can be told (TELL operator) on the domain of dicourse, and not by means of the implemented data structures. On the language level, it corresponds to an abstract data type with generic operations, which has been called the "knowledge level interface" by Brachman and Levesque [10, 30]. The abstract objects of the database correspond *directly* to expressions of the language on the conceptual level with the following properties[17]:

- there is a direct correspondence to the type system

- the language system has an object-oriented view on the database: objects have an identity, they contain an internal state and methods, sharing is possible

- persistency is orthogonal to type

- persistent processes are definable

- the language system offers a mechanism for modularization

A very interesting approach to the implementation of an integrated architecture is taken by PCLOS (Persistent CommonLisp Object System)[18], which is organized to allow the simultaneous use of multiple, different DBMS. The language and program layer is insulated from differences among the databases by an intermediate layer, a *virtual database*. This technique could very well serve to implement our proposal.

Although a programming language like LIFE is not itself a knowledge representation language, it is a toolbox which provides the basic constructs for representation schemata, a constraint system, a uniform abstraction concept, as well as basic inference and control mechanisms on a high abstraction level. The systems architecture we envision is sketched in fig. 3. The data base system is realized as a server which accessed by the language system through a knowledge level interface. We suggest to augment it by a reason maintenance module which records data dependencies. This imposes a two-level structure on the DBMS, since according to the present state of the art RMSs reside in primary memory.

References

[1] Ait-Kaci, H.: *Type Subsumption as a Model of Computation*. In: Kerschberg, L. (Ed.) [22], 1986, 115–139

[2] Ait-Kaci, H., Nasr., R.: *LOGIN: A Logic Programming Language with Built-In Inheritance*. Jl. Logic Programming, Vol. 3, 1986, 185–215

[3] Ait-Kaci, H., Lincoln, R.: *LIFE: A Natural Language for Natural Language*. MCC Technical Report ACA-ST-074-88, Austin, TX, 1988

[4] Ait-Kaci, H., Nasr, R.: *Integrating Logic and Functional Programming*. Lisp and Symbolic Computation, Vol. 2, 1989, 51–89

[5] Albano, A., Attardi, G.: *Issues in Data Base and Knowledge Base Integration*. In: Schmidt, J., Thanos, C. (Eds.) [39], 1989, 283–287

---

[17]a similar view is held by Albano [5] and in the JOSHUA system, Rowley [37]
[18]cf. Paepcke [36]

[6] Backofen, R.: *Integration von Funktionen, Relationen und Typen beim Sprachentwurf. Teil II: Attributterme und Relationen.* Diplomarbeit, IMMD VI, Universität Erlangen-Nürnberg, Erlangen, 1989

[7] Backofen, R., Euler, L., Görz, G.: *Towards the Integration of Functions, Relations and Types in an AI Programming Language.* To appear in: Proceedings of GWAI-90, Berlin: Springer (IFB), 1990

[8] Brachman, R, Levesque, H.: *The Tractabiblity of Subsumption in Frame-Based Description Languages.* Proceedings AAAI-84, Los Altos: Kaufmann, 1984

[9] Brachman, R., Gilbert, V., Levesque, H: *An Essential Hybrid Reasoning System: Knowledge and Symbol Level Accounts of KRYPTON.* Proceedings of IJCAI-85, Los Altos: Kaufmann, 1985, 532–539

[10] Brachman, R., Levesque, H.: *What Makes a Knowledge Base Knowledgeable? A View of Databases from the Knowledge Level.* In: Kerschberg, L. (Ed.) [22], 1986, 69–78

[11] Brodie, M., Mylopoulos, J.: *On Knowledge Base Management Systems.* New York, Springer, 1989

[12] Brodie, M., Manola, F.: *Database Management: A Survey.* In: Schmidt, J., Thanos, C. (Eds.) [39], 1989, 205–235

[13] Dessloch, S., Leick, F.-J., Mattos, N.M.: *An Approach to Knowledge Base Languages.* University of Karlsruhe, Computer Science Department, Karlsruhe, 1990; submitted for publication

[14] Dittrich, K.: *Objektorientierte Datenbanksysteme.* Informatik-Spektrum, Bd. 12, 1989, 215–220

[15] Euler, L.: *Integration von Funktionen, Relationen und Typen beim Sprachentwurf. Teil I: Konzeption, Typenhierarchie und Funktionen.* Diplomarbeit, IMMD VI, Universität Erlangen-Nürnberg, Erlangen, 1989

[16] Harel, D.: *Logic and Datbases: A Critique.* ACM SIGPLAN Notices, Vol. 22, No. 3, 1987, 14–20

[17] Haynes, C.T.: *Logic Continuations.* Proceedings of the Third Conference on Logic Programming, Berlin: Springer (LNCS 225), 1986, 671–685

[18] Hillis, D.: *New Computer Architectures: A Survey.* In: Brodie, Mylopoulos (Ed.) [11], 1989, 529-533

[19] Hillis, D.: *Parallel Computers for AI Databases.* In: Brodie, Mylopoulos (Ed.) [11], 1989, 551-563

[20] Israel, D.: *AI Knowledge Bases and Databases.* In: Brodie, Mylopoulos (Ed.) [11], 1989, 71–75

[21] Israel, D.: *Notes on Inference: A Somewhat Skewed Survey.* In: Brodie, Mylopoulos (Ed.) [11], 1989, 97–109

[22] Kerschberg, L. (Ed.): *Proceedings from the First International Conference on Expert Database Systems.* Redwood City, Ca.: Benjamin/ Cummings, 1986

[23] Lenat, D. et al: *KNOESPHERE: Building Expert Systems with Encyclopedic Knowledge.* Proceedings of IJCAI-83, Karlsruhe, Los Altos: Kaufmann, 1983, 167–169

[24] Lenat, D. et al.: *CYC: Using Common Sense Knowledge to Overcome Brittleness and Knowledge Acquisition Bottlenecks.* AI Magazine, Vol. 6, No. 4, 1986, 64–85

[25] Lenat, D., Feigenbaum E.: *On the Tresholds of Knowledge.* Proceedings of IJCAI-87, Milano, Los Altos: Kaufmann, 1987, 1173–1182

[26] Lenat, D.B., Guha, R.V.: *Buildung Large Knowledge-Based Systems — Representation and Inference in the Cyc Project.* Reading, Mass.: Addison-Wesley, 1989

[27] Levesque, H.: *A Fundamental Tradeoff in Knowledge Representation and Reasoning.* In: Brachman, R., Levesque, H. (Eds.): *Readings in Knowledge Representation.* Los Altos: Kaufmann, 1985, 42–70

[28] Levesque, H.: *Making Believers out of Computers.* Artificial Intelligence Journal, Vol. 30, 1986, 81–108

[29] Levesque, H.: *Knowledge Representation and Reasoning.* Annual Review of Computer Science, 1986, 1:255–287

[30] Levesque, H., Brachman, J.: *Knowledge Level Interfaces to Information Systems.* In: Brodie, Mylopoulos (Ed.) [11], 1989, 13–34

[31] Marr, D.: *Vision — A Computational Investigation into the Human Representation and Processing of Visual Information.* New York: Freeman, 1982

[32] Mattos, N.M., Michels, M.: *Modeling with KRISYS: The Design Process of DB Applications Revisited.* Proceesings of the 8th International Conference on the Entity-Relationship Approach, Toronto, 1988

[33] Morgenstern, M.: *The Role of Constraints in Databases, Expert Systems, and Knowledge Representation.* In: Kerschberg, L. (Ed.) [22], 1986, 351–368

[34] Nebel, B., Smolka, G.: *Representation and Reasoning with Attributive Descriptions.* IBM IWBS Report No. 81, Stuttgart, 1989

[35] Nebel, B: *Reasoning and Revision in Hybrid Representation Systems.* Berlin: Springer (LNCS 422), 1990

[36] Paepcke, A.: *PCLOS: A Critical Review.* Proceedings of the 1989 Conference on Object-Oriented Programming: Systems, Languages and Applications. SIGPLAN Notices, Vol. 24, No. 10, Oct. 1989, 221–237

[37] Rowley, S. et al.: *Joshua: Uniform Access to Heterogeneous Knowledge Structures or Why Joshing is Better than Conniving or Planning.* Proceedings of AAAI-87, Los Altos: Kaufmann, 1987, 48–52

[38] Schefe, P.: *On Definitional Processes in Knowledge Reconstruction Systems.* Proceedings of IJCAI-87, Milano, Los Altos: Kaufmann, 1987, 509–511

[39] Schmidt, J., Thanos, C. (Eds.): *Fundamentals of Knowledge Base Management Systems.* New York: Springer, 1989

[40] Shepherd, A., Kerschberg, L.: *Constraint Management in Expert Database Systems.* In: Kerschberg, L. (Ed.) [22], 1986, 309–331

[41] Smith, B.C.: *Reflection and Semantics in LISP.* Proceedings of the 1984 ACM Conference on Principles of Programming Languages, Salt Lake City, Utah, 1984, 23–35

[42] Sowa, J.: *Using a Lexicon of Canonical Graphs in a Semantic Interpreter.* In: Evens, M. (Ed.): *Relational Models of the Lexicon.* Cambridge: Cambridge University Press, 1988,113–137

[43] Stanfill, C., Waltz, D.: *Toward Memory-Based Reasoning.* Communications of the ACM, Vol. 29, 1986, 1213–1228

[44] Symbolics STATICE, Doc. No. 999702, Cambridge, Mass., 1988

[45] Weyer, S.: *The Design of a Dynamic Book for Information Search.* Int. J. Man-Machine Studies, Vol. 17, 1982, 87–107

[46] Weyer, S., Borning, A.: *A Prototype Electronic Encyclopedia.* ACM Transactions on Office Information Systems, Vol. 3, 1985, 63–88

[47] Zaniolo, C. et al: *Object Oriented Database Systems and Knowledge Systems.* In: Kerschberg, L. (Ed.) [22], 1986, 50–65

A KBMS for BABYLON

Conception and Development of a Knowledge Base Management System for a Hybrid Expert System Tool

Dieter Hovekamp

VW-GEDAS mbH
Pascalstraße 11
1000 Berlin 10

Michael Sandfuchs

ADV/ORGA AG
Kurt-Schumacher-Str. 241
2940 Wilhelmshaven

1. Abstract

Generally, expert system tools - whether being written in C, Prolog or Lisp - hold the processed knowledge in main memory. Usable expert systems rapidly require fast access to large amounts of knowlegde. A "Knowledge Base Management System" (KBMS) managing knowledge structures on a secondary storage is needed. As can be seen from many other contributions to this workshop today´s Data Base Management Systems (DBMS) are inadequate for this task because of knowledge being highly structured, manifold and varying. Therefore we believe that a KBMS should not be based on a DBMS; different techniques are required. This is why our field-manager Mr. Gottfried B. Bertram developed an idea of using specific set-oriented theoretical concepts in order to create a KBMS. At ADV/ORGA we developed an initial version of a KBMS which makes available the essential functions for knowledge base software systems like the AI-workbench BABYLON in the form of a C-library. The results represent a new approach to knowledge management on mass memory devices.

2. Introduction

Up to now expert systems hold and organize the entire knowledge which has been processed in main memory. Therefore large capacities of primary storage are needed, but generally are not available in commercial computer systems. In the course of working on the project WEREX the problem of large capacity main storage for AI-programs became ever more evident. This problem is being enhanced when processing sets of knowledge that are the size of present-day data bases and stored in mass memories [Brodie et al., 86].

Most desk top machines are not in command of virtual memory management techniques - a method frequently used by AI-machines. However, this technique is not a general solution to the problem of managing complex mass data because in general it only supports the running program and has no general interfaces to additional computing processes. Thus the use of large expert systems on desk top machines is being highly restricted. Using the main storage unit for administering complex knowledge structures is no general solution.

In addition to that the use of data base systems for storing knowledge is being hampered by conceptional problems. Present-day data-bases have been designed to administer large amounts of equally structured data. In contrast knowledge base systems comprise a wealth of structures and relatively few homogenous elements. Also data bases generally presuppose predefined descriptions of structures. In AI-systems these are not available - especially descriptions of dynamically produced structures. It is evident that these deficiencies being characteristic for present-day data base technologies require new concepts [Brodie et al. 86; Kerschberg 86; Friesen et al. 89]. This holds true for other areas as well (e. g. CASE, CIM or CAD [Dittrich et al., 86]).

Especially for the development of expert systems there is a lack of a tool that enables knowledge processing in mass memories similar to a data base management system. This would be a knowledge base management system (KBMS). Such a system would have to comprise a flexible and powerful form of knowledge representation which would allow for processing the instances of structures as well as the structures themselves.

In order to avoid the defiles in AI-applications mentioned above it is necessary to construct a system that enables processing and storing knowledge in secondary memories. To this end we worked out this concept and developed a first prototype within the framework of the WEREX-Project. This paper describes the theoretical concepts and first results.

We aimed to develop a KBMS in the form of an administration- and processing-system for representing knowledge, storage, access and manipulation on mass memories like hard disks and other media. The system organizes the representation on mass memory. It allows addressing knowledge structures on all levels as far down as the elementary level. It reads and writes in those knowledge units addressed directly in mass memory.

In this system knowledge is not stored in source form so that it is not necessary to restart or reinterpret the knowledge base. It is continiously possible to access knowledge structures without further preparation which also enables accesses not foreseen at the time of creation. Additionally it is possible to dynamically enhance the system in a enduring way.

Because of the flat structure of relational data bases not being sufficient for processing complex knowledge the KBMS comprises flexible forms of knowledge representation. Therefore KBMS exceeds relational data bases considering the aspects below:

- Data declaration (written in the data definition language) and contents (access via data manipulation language) are no longer separated. Structures and contents are jointly represented.

- It is possible to represent hierarchical structures and generic characterictics. Each element may be structured any further.

- When evaluating stored knowledge direct access on all levels as well as cross references are possible.

A prototype of KBMS has been developed and evaluated. In WEREX we designed concepts for its implementation into AI-tools especially BABYLON. The system BABYLON having a modular conception facilitates adding seperately developed parts on different levels. Thus a good starting point for realizing a prototypical use of this KBMS was given.

This realization is based on Extended Set Theory [Childs, 74; Bertram et al., 79]. Knowledge and facts as well as the relations among facts are represented by ordered sets . Knowledge is stored in a secondary memory (mass memory). Stored memory is being accessed by functions realized in the programming language C which guarantees a high portability to different hardware-environments and easy implementation into diverse software-systems.

3. Theoretical Foundations

3.1 Set-oriented Approach

Knowledge representation as it is realized in this KBMS is based on an enhanced form of set theory, which has been presented as a mathematical theory that is comprehensive and consistent, e. g. free of contradictions [Childs, 74; Childs, 78].

Looking at different forms of representing knowledge - i. e. semantic networks [Barr, Feigenbaum, 81], it becomes evident that relations between objects like "is a", "has a" or "is part of" are essential for the information it conveys. These relations between elements of knowledge express relations between the whole and its parts (part-whole-relation) or refer to specializations of more general objects.

In order to describe such relationships that are connected to a generic quality we shall introduce a generalized membership-relation for objects. This is the ϵ-relation (element of) which stands for the relations "is a", "has a" and "has property".

This common elementary relation points at using set theory as a starting point for an innovative form of representing knowledge which makes it possible to express all the relationships between objects mentioned above.

3.2 Representation of sets

3.2.1 Classical sets

Definition of sets by Cantor:

"A set M is the combination to a whole of certain well distinguished objects in our view or our thinking. These objects are called elements of M."

This is not a genuine definition that reduces a new concept to old ones. It illustrates an intuitive concept of set. If you follow a more precise concept of set you can describe the relationship between a set M and its elements using the elementary relationship ϵ as follows:

a ϵ M; this means that a is an element of the set M.

Set M can be described as follows:

a) by enumerating the objects (provided the number of objects is final),

b) by characterizing the objects in an unequivocal way.

Using method a) one simply writes down the signs for the (final number of) objects and puts them in parentheses. This is called an extension.

Example:

{ 1, 3, 5, 20 }

Describing objects according to method b) it is necessary to note the specific quality of an object in order to be part of a set. Describing objects by specifying a descriptive quality is called an intension.

Example:

The_first_three_natural_numbers   { 1, 2, 3 }

The following example is designed to illustrate the possibilities for structuring within a set representation using the e-relation:

birds { birds_of_prey, scavengers, singing_birds, ...}

This set can be structured in a more detailed way as follows:

```
birds {     birds_of_prey{ eagles      { ... }
                           owls        { ... }
                           falcons     { ... }}

            scavengers{    crows       { ... }
                           magpies     { ... }
                           ravens      { ... }}

            singing_birds{ finches     { ... }
                           titmice     { ... }
                           thrushes    { ... }}

            ... {             ...      { ... }}}
```

The intension "birds" includes the extension "birds_of_prey, scavengers, singing_birds, ...". The extension "birds_of_prey" is also an intension including the extension "eagles, owls, falcons, ...". Thus the element "birds_of_prey" is an extension related to the element "birds" as well as an intension related to the elements "eagles, owls, falcons, ...".

Intensions and extensions are combined and represented in connection with their relations in a set thereby facilating a common access to intension and extension.

It is possible to retrieve elements or structured elements respectively of the sets on the basis of their attributes or qualities (element relationship). All elements (objects) belonging to the set "birds" can be written as "birds (x)", "x" marking a space for all elements possible carrying the denoted quality.

Below there are some possible queries (Q) as well as the corresponding results (R):

Q: birds (x)
R: birds { birds_of_prey, scavengers, singing_birds, ... }

Q: birds (birds_of_prey(x))
R: birds { birds_of_prey{ eagles, owls, falcons}}

Q: y (x(crows(rook)))
R: birds { scavengers{ crows{ rook }}}

Q: x (y(blackbirds))
R: singing_birds{ thrushes{ blackbirds}}

Within a set bracket all elements have equal rights. They have the same quality. In view of the set structure they are on the same level.

3.2.2 Extended sets (n-tupel)

N-ary relations cannot be represented by a simple e-relation or single predicates as described in chapter 3.2.1.

Example: River <flows through> a_country.

x	y
Rhone	France
Rhine	Switzerland
Main	Germany
Rhine	Germany
...	

The binary relation "flows_through (x,y)" cannot be represented by "element_of" or the "included_in"-relation in simple sets. Both the elements x and y of the relation are contained in the set "flows_through". However, they are not exchangeable having a different meaning (name of the river, name of the country). In the example above "x" stands for "river" and "y" stands for "country". Exchanging the elements would for instance result in "Switzerland flows through Rhine".

It is necessary to introduce n-ary relations for more complex relationships. This can be done by using n-Tupels (pairs, triples, quadruples, ...). These are ordered sets in which the position of an element or the order of the elements respectively is important. Using the example "River flows through country" this means:

flows_through $\{ x^1, y^2 \}$ includes the sets

$$\{ Rhone^1 \quad France^2 \}$$
$$\{ Rhine^1 \quad Switzerland^2 \}$$
$$\{ Main^1 \quad Germany^2 \}$$
$$\{ Rhine^1 \quad Germany^2 \}$$
...

The index numbers put up point to the respective position of the element within the set. The position in the tupel indicates the role, that an element plays within a relation (scope of definition, scope of values, attributes).

In each extension objects are being listed. In their respective extension they can be ordered at will and structured any further. Each extension may have an intension assigned to itself which names or describes the enlisted objects.

Classical sets (see 3.2.1) may be regarded as a special case of ordered sets. Here all elements are in position "1". Thus n-tupels enable us to describe ordered as well as classical sets. This form of enhanced classical representation of sets is called Extended Set Theory (XSet-Theory) [Childs, 74].

XSet-Theory introduces the membership-relation by positioning the elements.

$a\ e_i\ A$ means: a is an element of A in position i.

Unordered sets (see above) are relations with all elements being in the first position.

Examples of relations:

single relations
$$A\{a^1\}$$
$$\{\ 1^1, 2^1, 3^1\ \}$$

Pairs
$$P\{a^1, b^2\}$$

Triples
$$T\{a^1, b^2, c^3\}$$

Quadruples
$$Q\{a^1, b^2, c^3, d^4\}$$

Examples of n-tupels:

In the tupel $\{a^1, b^2, d^4\}$ position 3 has not been taken.

In the tupel $\{a^1, b^1, d^2\}$ position 1 has been taken twice.

The tupel $\{a^1, b^1, c^1\}$ represents the classical set $\{a,b,c\}$.

These few examples illustrate the versatility of the representation by n-tupels (see 3.4).

A classical data base set may also be represented like $\{a^1, b^2, c^3, d^4\}$ with field 1 corresponding to "a", field 2 corresponding to "b" and so on. However, in a DBMS there is no possibility to further structure an element with simple means (e.g. structuring "b" in position 2). In general it becomes necessary to redefine (or newly define) the data base.

Using a relational DB the following example is cumbersome to realize:

an expression/a relation family

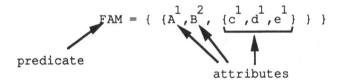

is not put down as

FAM

F	M	C
A	B	c,d,e
C	D	–

but in its first normal form

F	M	C
A	B	c
A	B	d
A	B	e
C	D	–

3.3 Pure intensions in knowledge representation

In addition to those set representations introduced so far via XSets additional representations are needed that consist of only one intensional representation. In particular these are agreements concerning different viewpoints on "knowledge bases", cross references to other elements and knowledge bases.

Within a knowledge base pure intensions serve as rules for obtaining extensions. If there is a pure intension when analysing an inquiry you gain access to a different knowledge base or a different part of the topical knowledge base respectively via the intension. Furthermore they are used for alias names which may be used for defining different views on the knowledge base (user orientation). By combining intensions it is possible to structure the knowledge base in a different way and also to build up new ones.

Examples of pure intensions:

$$\text{Family}^1\{ \text{ Smith}^1\{ \text{ Joseph}^1, \text{ Mary}^2, \{ \text{ Jack}^1, \text{ Liesel}^1 \}^3\}$$

$$\text{Miller}^1\{ \text{ Fritz}^1, \text{ Eva}^2, \quad \{ \text{ Uwe}^1 \}^3\}$$

$$\text{Jansen}^1\{ \text{ Adam}^1, \text{ Frieda}^2, \{ \text{ Jack}^1 \}^3\}$$

$$\text{etc...}\{ \quad \text{...} \quad\quad\quad \{ \text{ ... } \}\}\}$$

family names :

$$\text{Family}^1 (\text{ X}^1)$$

families with father "Jack":

$$(\text{ X}^1 (\text{ Jack}^1))$$

father :

$$\text{Family}^1 ((\text{ X}^1)^1)$$

mother :

$$\text{Family}^1 \ ((\ x^2\ )^1)$$

childs :

$$\text{Family}^1 \ (((\ x^1\ )^3)^1)$$

childs with name of family:

$$\text{Family}^1 \ (\ y^1((\ x^1\ )^3))$$

You gain access to the knowledge base via intensions that contain variables:

- $\text{Family}^1 (X^1)$ produces

  $$\text{Family}^1 \ \{\text{Miller}^1, \ \text{Meier}^1, \ \ldots\}$$

- $\text{Family}^1((X^1, Y^2)^1)$ produces

  $$\text{Family}^1 \ \{\{\text{Fritz}^1, \ \text{Eva}^2\}^1, \ \{\text{Paul}^1, \ \text{Erna}^2\}^1\}$$

This may be used to define new intensions:

- married $:= \text{Family}^1((X^1, Y^2)^1)$

- childs $:= \text{Family}^1((X^1)^1)$

During access the intensions can be combined by logical operations.

Intensional representations are needed as references to extensions if the elements of the extensions reoccur in other places of the knowledge base, e.g. in a different context. Intensions being non-procedural access rules are required to be storable as elements in extensions.

Example: lives_in $\{\ \{\text{Cologne}^1, \ \text{Smith}\{\ \}^2\}\}$.

In this example there is an intension without an extension in position two. Via this intension access is gained to a different part of the knowledge base or to a different knowledge base.

3.4 Set-oriented operations

In this KBMS knowledge structures are represented as (extended) sets. In order to realize processing operations on knowledge structures one may make use of set theoretical operations. Above all these include "union", "intersection", "set difference" and the quantifiers "existential-quantifier", "universal-quantifier" ($\exists, \forall$).

The following set theoretical operations are realized:

union	$A \cup B$	
intersection	$A \cap B$	
set difference	$A \setminus B$	(elements of set A without the elements of set B)
existential	$\exists\, x \in A$	
universal	$\forall\, x \in A$	

The meaning of these set theoretical operations becomes evident in view of the update-function:

$A = \{a^1, b^2, c^3, d^4\}$ be a partial structure of a knowledge base (set of data); element c in position 3 is to be replaced by element w (update).

Set $B = \{c^3, w^3\}$ includes the elements to be exchanged.

Operations A\B and B\A produce the following sets:

$$A \setminus B = \{a^1, b^2, d^4\} \qquad\qquad B \setminus A = \{w^3\}$$

The combination of both difference sets results in:

$$(A \setminus B)\ \cup\ (B \setminus A) = \{a^1, b^2, d^4\} \cup \{w^3\} = \{a^1, b^2, w^3, d^4\}$$

Thus element c^3 has been effectually exchanged by element w^3 in the partial structure of the knowledge base.

It is possible to combine the results of inquiries or pure intensions respectively in a set theoretical way provided they can be sorted with the help of sorter or merge-programs (see 4.5).

4. Implementation

4.1 Software conception

In order to use the KBMS presented here for processing and storing knowledge in secondary mass memories several levels of description have to be considered. We distinguish between a user-oriented presentation (i. e. a window-oriented browser), a text presentation and an internal presentation .

The graph below shows the storing process and the analization of inquiries as it applies to a knowledge base.

Levels of
representation

User-
representation

Text-
representation

Internal
representation

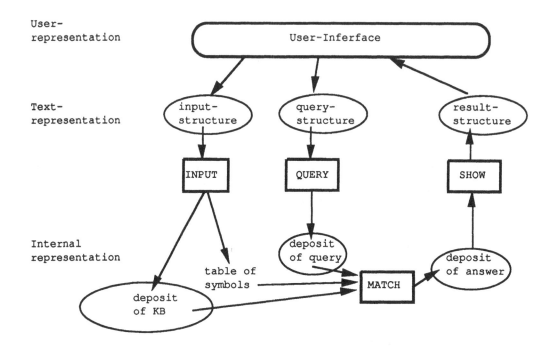

input structure	knowledge base as text representation
query structure	query to knowledge base as text representation
deposit of KB	knowledge base in internal representation
table of symbols	file representation of symbol addresses
query deposit	query to knowledge base internal representation
result deposit	internal result for the query to knowledge base
result structure	result of query as text representation

The levels of representation in detail:

User representation

Within the WEREX-project we developed a prototype of a KBMS that comprises the representation of text and internals. It was not intended within this project to devolop a user oriented representation by implementing the respective utilities like a graphic editor, a browser etc. In order to be able to use the query and access functions we developed an ASCII-surface which makes handling easy and clear.

Text representation

As a matter of simplicity we presume that the knowledge base (input structure) as well as possible queries (query structure) for the knowledge base are realized in a text file. In these files the elements and their interrelationships are represented legibly. This form of representation corresponds to the formal syntax introduced in the chapter above. The syntax will be described in chapter 4.3. The results (structure of results) of queries or accesses respectively addressed to a knowledge base will be processed correspondingly.

Internal representation

In order to store and process knowledge a form of representation (internal representation) will be selected that is compatible to the memory medium. The internal description (see chapter 4.4.2 for more details) contains all elements and their interrelationships - as does the text description. The internal description is fitted to the memory medium and avoids the disadvantages of a linear storage system by comprising redundant information.

Transformation between text representation and internal representation

The functions "INPUT" and "QUERY" are being used to transform the text representation into an internal representation. The function "INPUT" transforms the input structure into a representational data set (stored as a file) which comprises all elements and their interrelationships. The -relations of all elements are stored as a pattern. In addition to that a table of symbols is being produced which enables fast access to the elements of the storage data set. The function "QUERY" also exchanges the query structure into a storage data set (deposit of query). This produces the same patterns as those in the representational data set, thus making use of comparisons of patterns for searching in the representational data set.

After transformation the knowledge base and the query exist in the form of three files in the secondary memory. The function "MATCH" carries out the comparison of patterns (pattern-matching). With the help of the query "MATCH" navigates within the knowledge base stored in secondary memory and extricates all those elements out of the representational data set that match the query. The result of this query is represented in the form of an result deposit. This resulting data set contains that part of the knowledge base which matches the query and therefore is structured like a representational data set.

The transformation of an result data set into the text representation is done by the function "SHOW".

At a later point it will be possible to develop functions for the interface into a window system. This will allow a user-friendly representation and graphical processing of knowledge.

4.2 Software and examples

The prototypical realization of KBMS comprises four different programs for transforming the knowledge base, transforming the query, querying the knowledge base and representing the results.

Additionally there are examples of knowledge bases belonging to the areas of classification and configuration.

4.3 Agreements on structure

Concerning the text representation of a knowledge base and its input structure and query structure we suggest the following:

4.3.1 Input structure

In a knowledge base (input structure) there are three types of elements:

intensions with a subsequent structure

    <position number>, ":", <symbol>, "{", <structure>,"}"

extensions without structure (atoms)

    <position number>, ":", <symbol>

anonymous intensions

    <position number>, ":", "{",<structure>,"}"

Example "family":

```
1:family{ 1:smith{ 1:joseph
                   2:mary
                   3:{ 1:jack 1:liesel }
                   4:{ 1:wilhelm 2:martha }
                   5:{ 1:jack 2:rita }
                   }
         }
```

The character ":" is being used to mark the separation between the position number and the symbol. The position number and the separation mark are always present. A deeper structuring of an element is started by "{" and ended by "}". The symbol exists of alphanumeric characters, the first of which has to be a letter. An element like e. g. "1:joseph" which has no further structure is called an atom. In the case of an anonymous structure "3:{...}" there is no symbol after the colon.

Those elements (e.g. numerics) that are not meant to be incorporated into the table of symbols will be marked as anonymous. This being the case they are advanced by "#". Thus these elements cannot be used for queries.

4.3.2 Query structure

Query structure and input structure are built up correspondingly except for the following differences:

There are no anonymous elements. Those variables for which we want to collect elements during the query start with an underscore '_'. A further structuring of an element is started by "(" and ended by ")".

Like in the input structure there are three cases in the query structure:

Intensions with a following structure

    <position number>, ":", <symbol>, "(", <structure>, ")"

extensions without a structure (atoms)

    <position number>, ":", <symbol>

anonymous intensions

    <position number>, ":", "(", <structure>,")"

Example 'family':

names of father and mother of all families named Smith

```
1:( 1:smith( 1:_x 2:_y ))
```


4.3.3 Description of syntax

In the legible representation the following agreements apply to the syntax.

1:<symbol>{ .... }	extension, intension
1:{ .... }	extension (anonymous)
1:<atom>	extension, intension
2:#<symbol>	anonymous extension

For knowledge representation we use the following syntactic marks:

"(" beginning of a structure (query)

")" end of a structure (query)

"{" beginning of a structure (representation)

"}" end of a structure (representation)

"#" anonomized element following

"_" variable following

":" seperator between position number and symbol

All other notations like Blank, CRLF, Tab, FF are processed as element-seperators. EOF marks the end of the data set.

4.4 Structure of a data set

We shall use the knowledge base 'family' as an illustrating example for the structure of a data set. The data sets used by the system are implemented as files and discribed in the following order:
- legible representation of the knowledge base
- internal representation of the knowledge base with table of symbols
- legible representation of a query
- internal representation of a query
- result of query in the internal form
- result in legible representation

4.4.1 Text representation of the knowledge base

The text representation as it is presented in the following example "family" is basic for the construction of a knowledge base:

```
1:family{ 1:smith{  1:joseph
                    2:mary
                    3:{ 1:jack 1:liesel }
                    4:{ 1:wilhelm 2:martha }
                    5:{ 1:jack 2:rita }
                  }
          1:miller{ 1:fritz
                    2:eva
                    3:{ 1:uwe }
                    4:{ 1:hugo 2:renate}
                  }
          1:miller{ 1:fritz
                    2:carla
                    3:{ 1:udo 1:horst }
                    5:{ 1:rudolf 2:beate }
                  }
          1:jansen{ 1:adam
                    2:frieda
                    3:{ 1:jack 1:christiane 1:fritz
                        1:susanne }
                    4:{ 1:helmut 2:erna }
                    5:{ 1:frank 2:karin }
                  }
        }
```

4.4.2 Internal representation of a knowledge base

The legible representation of the knowledge base is represented in a representational data set and a table of symbols. This form of representation (internal representation) is adapted to the storage medium and balances out the disadvantages of a linear memory by means of information redundancy.

The representational data set contains all pieces of information (elements and their interrelationships). The table of symbols contains all symbols (intensions, extensions) that are not anonymous or anonymized. This allows efficient adressing within the representational data set.

The representational data set contains all structure names and atoms. Structures without a name receive '@' as a dummy-name. The elements of an entry are separated by a TAB-sign. The inputs are separated by a LineFeed. An input consists of:

- name of element

- level of nesting

and for each level:

- serial number

- position

Element names are corresponding to the symbols of the input structure. The level of nesting indicates the structure depth of the respective element. The serial number helps to distinguish between elements that belong to the same intension (e.g. family name). The position marks the role of the element (relation in the tupel).

Each element inherits the pieces of information (serial number and position) of the corresponding intension. For instance: 'Smith' inherits the serial number '1' and the position '1' from 'family'. 'Joseph' and 'Mary' inherit all pieces of information from 'Smith'. The inputs 'Joseph' and 'Mary' have different final serial numbers and positions.

```
structure-name
  |          /--------------------------  level of nesting
  |          |  /-----------------------  1   serial number
  |          |  |  /--------------------  1   position
  |          |  |  |  /-----------------  2   serial number
  |          |  |  |  |  /--------------  2   position
  |          |  |  |  |  |  /-----------  3   serial number
  |          |  |  |  |  |  |  /--        3   position
  |          |  |  |  |  |  |  |
family     0  1  1
smith      1  1  1  1  1
joseph     2  1  1  1  1  1  1
mary       2  1  1  1  1  2  2
@          2  1  1  1  1  3  3
jack       3  1  1  1  1  3  3  1  1
liesel     3  1  1  1  1  3  3  2  1
@          2  1  1  1  1  4  4
wilhelm    3  1  1  1  1  4  4  1  1
martha     3  1  1  1  1  4  4  2  2
@          2  1  1  1  1  5  5
jack       3  1  1  1  1  5  5  1  1
rita       3  1  1  1  1  5  5  2  2
miller     1  1  1  2  1
fritz      2  1  1  2  1  1  1
eva        2  1  1  2  1  2  2
@          2  1  1  2  1  3  3
uwe        3  1  1  2  1  3  3  1  1
@          2  1  1  2  1  4  4
hugo       3  1  1  2  1  4  4  1  1
renate     3  1  1  2  1  4  4  2  2
miller     1  1  1  3  1
fritz      2  1  1  3  1  1  1
carla      2  1  1  3  1  2  2
@          2  1  1  3  1  3  3
udo        3  1  1  3  1  3  3  1  1
horst      3  1  1  3  1  3  3  2  1
@          2  1  1  3  1  4  5
rudolf     3  1  1  3  1  4  5  1  1
beate      3  1  1  3  1  4  5  2  2
jansen     1  1  1  4  1
adam       2  1  1  4  1  1  1
frieda     2  1  1  4  1  2  2
@          2  1  1  4  1  3  3
jack       3  1  1  4  1  3  3  1  1
christiane 3  1  1  4  1  3  3  2  1
fritz      3  1  1  4  1  3  3  3  1
susanne    3  1  1  4  1  3  3  4  1
@          2  1  1  4  1  4  4
helmut     3  1  1  4  1  4  4  1  1
erna       3  1  1  4  1  4  4  2  2
@          2  1  1  4  1  5  5
frank      3  1  1  4  1  5  5  1  1
karin      3  1  1  4  1  5  5  2  2
```

In order to have faster access to the representational data set the table of symbols should contain the cross references in sorted sequences.

All elements of an input are divided by a TAB-sign. All inputs are separated by a LineFeed. An input consists of:

- element name

- absolute byte-position in the REP-file (= its Address).

Example:

element-name	byte-position in REP-file
family	0
smith	15
joseph	34
mary	55
jack	93
liesel	117
wilhelm	160
martha	187
jack	230
rita	254
miller	278
fritz	297
eva	318
uwe	354
hugo	394
renate	418
miller	444
fritz	463
carla	484
udo	522
horst	545
rudolf	587
beate	613
jansen	638
adam	656
frieda	676
jack	715
christiane	739
fritz	769
susanne	794
helmut	838
erna	864
frank	905
karin	930

4.4.3 Text representation of a query

The text representation of a query has a similar structure to the text representation of a knowledge base. Variables are commenced by an underscore.

Example: Name of father and mother of all families called Smith

```
1:( 1:smith( 1:_x 2:_y ))
```

The name "smith" serves as an anchor. The information contained in the top level is not wanted (dummy-input). _x serves as an empty space (variable) for "father", _y for "mother".

This text representation of a query serves for producing an internal representation as it is the case for knowledge bases, too.

4.4.4 Internal representation of the query

The internal representation of a query is structured similar to the representational data set. All elements of an input are divided by a TAB-sign. All inputs are seperated by a LineFeed. An input consists of:

- element name,

- level of nesting,

and for each level:

- position.

The element names are the same as the symbols used in the query structure. The level of nesting indicates the structure depth of the respective element. Serial numbers as in representational data sets are not needed. The position marks the role, the element plays (relation in tupel). Any one position within one level may not be taken several times. There has to be at least one intension (anchor) within a query structure in order to gain an access point into the representational data set via the table of symbols.

The following query data set is based on the query presented in chapter 3.4.3:

```
structure-name
   |                /------------------ level of nesting
   |               |   /-------------- 1   position
   |               |  |   /---------- 2   position
   |               |  |  |   /------ 3   position
   |               |  |  |  |
   |               |  |  |  |
@                  0  1
smith              1  1  1
_x                 2  1  1  1
_y                 2  1  1  2
```

4.4.5 Result of query in the internal form

Initially the result of a query delivered by the 'MATCH'-program exists in the form of a representational data set (see 4.4.2). This form of representation enables further processing of the result (for set theoretical operations see chapter 4.6).

The query presented in the last chapter has the following result:

```
structure-name
   |          /--------------------------- level of nesting
   |          |    /--------------------- 1   serial number
   |          |    |   /----------------- 1   position
   |          |    |   |   /------------- 2   serial number
   |          |    |   |   |   /--------- 2   position
   |          |    |   |   |   |   /----- 3   serial number
   |          |    |   |   |   |   |  /- 3   position
   |          |    |   |   |   |   |  |
   @          0    1   1
   smith      1    1   1   1   1
   joseph     2    1   1   1   1   1   1
   mary       2    1   1   1   1   2   2
```

4.4.6 Result in text representation

The conversion function 'SHOW' serves at presenting the result of the query in a legible representational form. After transforming the internal representation into the text representation the result of the query is as follows:

```
1: {
        1:smith {
                1:joseph
                2:mary
        }}
```

The indentations help to set out optically the element relations and the level of nesting of each element.

4.5 Set theoretical operations

All important functions for a KBMS like for instance Update may be realized with the help of the set theoretical operations "union", "intersection" and "set difference".

Those three operations can be remodeled by a merge-function, if the merge-inputs are available in sorted order. A sorted order of the results (match1 and match2) becomes possible because of the information-contents of each element (serial number and position) being usable as an unambiguous key.

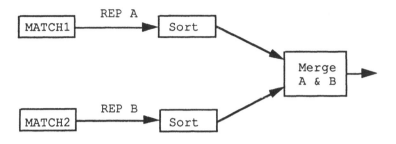

In order to set theoretically combine two input-streams by means of a merge-function it is necessary to have the contents of the input files in a sorted sequence. The operations "intersection", "union" and "set difference" can be realized via comparing the elements (analyzing the unambiguous key by means of <, =, >). The sets of results that come into existence via these operations are sorted automatically. They as well may be used as input for further set theoretical operations.

set operations (merge-function):

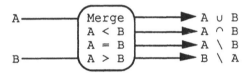

The following table shows the relationship between the operations and the logical comparisons. In this table the letters A and B stand for the elements that are contained in the sets (data sets) A and B.

	A ∪ B	A ∩ B	A \ B	B \ A	read next
A < B	A		A		A
A = B	A	A			A, B
A > B	B			B	B

- *set union*

A unification means that all elements of sets A and B are taken over into the set of results. Elements occurring in both sets are considered only once (compare the case A=B).

A<B: The element of set A is being put into the set of results; the next element of A is being read afterwards.

A=B: he element of set A is being put into the set of results; the next elements of A and B are being read afterwards.

A>B: The element of set B is being put into the set of results and the next element of B is being read afterwards.

- *intersection*

Only those elements belonging to both sets (A & B) are being incorporated into the set of results (compare case: A=B).

A<B: The element of set A is not being put into the set of results as it is not contained in B. The next element of A is being read afterwards.

A=B: The element of set A is being put into the set of results and the next elements of A and B are being read afterwards.

A>B: The element of set B is not being put into the set of results as it is not contained in A. The next element of B is being read afterwards.

- *set difference*

The set of results A \ B contains all elements of set A that are not contained in set B. Elements occurring in both sets are not being considered (compare case: A=B).

A<B: The element of set A is being put into the set of results as is not contained in B. The next element of A is being read afterwards.

A=B: The element of set A is not being put into the set of results and the next elements of A and B are being read afterwards.

A>B: The element of set A can not be processed until the element of set B is bigger than or equals the element of set A. The next element of A is being read afterwards.

The difference B \ A is formed correspondingly.

5. Integration within the hybrid expert system tool BABYLON

In the context of the project WEREX we explored the possibilities for integrating KBMS into XPS-development tools. We especially emphasized a conceptualization of realizing it with BABYLON.

Having a modular conception the AI-workbench BABYLON facilitates the incorporation of seperately developed software components on different levels. This is an ideal starting point for incorporating the KBMS presented here.

5.1 Possibilities of Integration

It is advisable to order the different fields of application of a KBMS according to the degree of connection with the tool-system. This ordering proceeds on the analogy of the distinction between loose vs. tight coupling of a data base with an expert system [de Buhr, 1988].

Under the most general circumstances the KBMS can be integrated into the BABYLON-system as a self-supporting processor. This would put a further highly flexible form of knowledge representation at the knowledge engineer's disposal.

A simpler version of incorporating KBMS is facilitated by using it to build up a case data administration of XPS-consultations. This link-up being relatively loose so that the protocols of single consultations would be held in the main memory and - with the help of KBMS - be filed in secondary memory only after terminating a session. This shows the big advantage of KBMS over conventional database: it is mainly the highly diversified possibilities of continued processing that successively built up stock of knowledge. Thus KBMS offers much more flexible possibilities for storing and continued processing than classic databases - even when using this simple form of link-up.

The direct representation of single BABYLON-processors on storing and processing mechanisms of the KBMS would constitute an especially tight link-up between BABYLON and KBMS. As a concrete example one could imagine a base for the frame-processor.

Finally it would be possible - and this holds true especially for BABYLON - to replace part of the functionality of the flavor system which is being used as a basis for implementation - by the KBMS. This would facilitate the storing of any object structures and the support of a common base-administration for all processors.

Initially this KBMS was realized at ADV/ORGA on personal computers. By using widespread programming languages (C and Lisp) it is possible to later transfer it to other computer systems as used in the WEREX-project.

5.2 Preview on the practical application of KBMS

The KBMS offers the prerequisites for knowledge processing in mass memories in all the link-ups thought of. Particularly this concerns the storage- and search-algorithms needed in the form of library routines.

A link-up with further forms of representations present in the system is being realized by means of those methods that the BABYLON-system offers to this end.

It remains to work out concepts about storing different forms of knowledge representation in a KBMS. This necessitates research as to how the realized version of a KBMS can administer and process data efficiently.

6. Summary

The Knowledge Base Management System presented here administers knowledge in secondary memories like e.g. harddisks; it organizes the storing of knowledge in the storage medium. Direct access to the knowledge base and cross references on all levels are feasable.

In a set theoretical approach this KBMS is based on making use of a general membership-relation and a standardization of the definition of classic sets and n-tupels referring to Child's 'extended set theory' [Childs, 74].

Apart from the so-called XSets comprising the classic set term as well as 'ordered sets' (n-tupels) the main feature of the proposed form of knowledge representation is the common representation of extensions and intensions. There may also be anonymous extensions, i. e. extensions that have no intensions or no higher order term respectively. The elements of an extension or an intension may be structured at will within the XSet-rules. Intensions without extensions are "predicates"; knowledge elements filling out these predicates are the instantiations of such an intension. XSets - be them complete or anonymous extensions or even pure intensions - are elements of a knowledge base.

Access and assignment are realized using intensions. An access is independent of the structure depth and produces a set that includes all elements fulfilling the requirements of the respective intension. This set of results may be empty, may contain one or more elements or partial structures of the knowledge base.

It is possible to do set theoretical operations on XSets. They are of practical importance for instance to implement an update-function. Access definitions and link-ups facilitate structural transformations of the knowledge base. The structure of the knowledge base is formalized and covers a wide field of knowledge representation.

The presented KBMS allows for statements on elements, attributes of objects and relations between elements. By means of this KBMS the distinction between declaration and contents, structure and instance, DDL and DML as well as definition and interpretation respectively is abolished. Subjects of processing are the elements and their structure.

Being very flexible the chosen approach facilitates the administration of those forms of representation that have so far been used in expert systems like rule sets, frames and semantic networks.

Acknowledgements

The ideas presented here would not have come into existence without the intensive work of Mr. Gottfried Bertram. He originated the concept, enhanced the scientific foundations and developed preliminary implementations of this set-oriented approach for a KBMS. We have highly profited from Mr. Bertram´s enthusiastic way of introducing us to his ideas and would like to express our special thanks to him. We would also like to thank him, Mr. Edzard de Buhr and all the other members of the AI-department at ADV/ORGA for their extremely useful cooperation. Additional thanks to Dr. Thomas Christaller and Ministerialrat Isensee for their coordination of the WEREX-project which this work was supported in by the Bundesministerium für Forschung und Technologie under contract number ITW8505B5.

References

[Barr, Feigenbaum 81] Barr A., Feigenbaum E.A.: *Knowledge Representation*, in Barr A. & Feigenbaum E.A.(Eds.):"The Handbook of Artificial Intelligence Volume 1, Chapter III". The William Kaufmann, Inc. Los Altos, California, 1981.

[Bertram et al. 79] Bertram G., Braig A., Gweißner S., Mandrella R., Matheis U.: *Mathematische Implementierungsmethoden für ein mengentheoretisches Datenbanksystem.* BMFT-Forschungsbericht DV 79-02, Battelle-Institut e.V. Frankfurt am Main, 1979.

[Braig et al. 81] Braig A., Gwießner S., Mandrella R., Matheis U., Welter J.: *Realisierung und Pilotanwendung eines Datenbanksystems auf mengentheoretischer Grundlage.* BMFT-Forschungsbericht DV 81-004, Battelle-Institut e.V. Frankfurt am Main 1981.

[Brodie et al. 86] Brodie M. L., Mylopoulos J. (Eds.): *On Knowledge Base Management Systems.* R.R. Donnelley & Sons, Harrisonburg, Virgina, 1986.

[de Buhr 89] de Buhr E.: *Datenbankanschluß für BABYLON auf dem IBM/PC*, in Christaller T., Di Primio F., Voss A. (Eds.): Die KI-Werkbank BABYLON, Addison Wesley, Bonn, 1989.

[Childs 74] Childs D.L.: *Extended Set Theory: A General Model for Very Large, Distributed, Backend Information Systems.* presented at the "Third International Conference on Very Large Databases", Tokio, Japan 1978.

[Childs 78] Childs D.L.: *Extended Set Theory: A Formalism for the Design, Implementation and Operation of Information Systems.* Set Theoretic Information System Corp., Ann Arbor Michigan, 1974.

[Dittrich 89] Dittrich, K.R.: *Objektorientierte Datenbanksysteme*, Informatik-Spektrum, (1989) 12, 215-220.

[Dittrich et al. 86] Dittrich K.R., Gotthard W., Lockemann P.C.: *DAMOKLES - A Database System for Software Engineering Environments.* Proceedings IFIP Workshop on Advanced Programming Environments Trondheim (Norway), 1986, Springer, Heidelberg, 1986

[Friesen et al. 89] Friesen O.Golshani F.: *Databases in Large AI Systems*, AAAI AI-Magazine Winter 1989.

[Kerschberg 86] Kerschberg L. (Ed): *Expert Database Systems - Proceedings From the First International Workshop.* The Benjamin/Cummings Publishing Company, Inc. Menlo Park, California, 1986.

Database Concepts for the Support of Knowledge-Based Systems

Thomas Ludwig   Bernd Walter *
University of Trier   Department of Mathematics and Computer Science
P.O. Box 3825 55   5500 Trier   Federal Republic of Germany

Abstract

The purpose of a database management system is the reliable and efficient management of shared persistent data. Current knowledge-based systems are designed for representing and processing knowledge in a way that is cognitively adequate. Since knowledge should be kept persistent as well, it seems to be interesting to combine the techniques of database management systems and knowledge-based systems into something like a knowledge base management system. We shortly show some conceptual differences between these two types of systems and then discuss how database concepts should be evolved in order to meet the requirements of knowledge-based systems. Using LILOG-DB as an example, we demonstrate in detail the modeling capabilities a database system should provide in order to support typical knowledge structures.

Especially, we discuss the design of *FLL*, a first-order logic based query language supporting an order-sorted complex-object data-model under an Open-World Assumption.

1   Introduction

What is knowledge, what is data ?

Up to now, there is no generally accepted definition of what *knowledge* is. Some view it as being a "symbolic representation of aspects of some named universe of discourse" ([14]), others define knowledge in a more abstract way, i.e. as being independent of a fixed model ([37]). From a human viewpoint, the second definition seems to be preferable, since it takes into account that knowledge is something which is not yet perfectly understood. However, in order to make a computing system knowledgeable (intelligent) it needs some symbolic representation of the needed knowledge. From the outside this representation is just a model of the real knowledge, but for the computing system itself, this representation is its real knowledge. So, depending on the viewpoint both definitions can be justified.

A *knowledge-based system* is a system that incorporates a symbolic representation of knowledge. However, how much a knowledge-based system really "knows" (how "intelligent" it is) depends on how flexible it is in interpreting and applying its local representation.

The type of symbolic representation and the mapping from knowledge to representation is determined by the selected knowledge representation method. In Artificial Intelligence knowledge representation has become one of the mayor topics of research.

A *database* is a "logically coherent collection of related data" ([13]). According to [1] a database is also a "model of an evolving physical world. The state of this model, at a given instant, represents the knowledge it has aquired from this world". Additionally, in [13] data are defined as "known facts". As we can see, "data" is used inconsistently as well, partly to denote an abstract concept ("known facts"), partly to denote a representation of this abstract concept. However, this inconsistent use of "data" can be explained in the same way as in the case of "knowledge". Furthermore these statements help us to relate "data" and "knowledge" by allowing us to state:

Data is a simple kind of knowledge.

In currently available database systems, the symbolic representation of the universe of discourse is simpler and less polished than in knowledge representation. In fact, in state-of-the-art data modeling many aspects

---

*The work reported here was carried out within the LILOG-project of IBM Germany

of the universe of discourse cannot be represented at all, i.e. database systems know nothing about defaults, indefinite facts, uncertainty and the like.

The term *knowledge* is more related to the human mind and reflects the objectives of artificial intelligence to duplicate human capabilities, whereas the term *data* is more related to traditional alpha-numeric computing and reflects the objectives of computer science of doing things efficiently. So, concerning the primary design goals we can state:

Knowledge representation means cognitive adequacy, data modeling means efficiency.

Data models might nevertheless include implementation independence as well as semantic concepts, but anything significantly decreasing efficiency is usually excluded. Therefore, most semantic data models only support concepts that can easily be mapped onto the relational model of data.

The early data models like the *network model* or the *hierarchical model* were primarily designed around implementation concepts and typical concepts of imperative programming languages. Most of nowadays data models are designed around mathematical concepts. The currently dominating *relational model* is based on first order predicate logic, with the database being a collection of sets of function free ground facts and queries being function free first order formulas. In the newly developed deductive database systems following the *DATALOG model*, a database consists of sets of horn-clauses (sometimes even functions are supported) whereas queries are just goals (as in PROLOG) or complete logic programs.

Knowledge representation was long dominated by informal versions of the frame and semantic net concepts. However, in order to better support formal reasoning, mathematical concepts became more important. Consequently, many new knowledge representation techniques have been formalized using first order predicate logic. So the state of the art can be characterized by following statement:

The formal foundation of both knowledge representation and data modeling lies in first order predicate logic and several "extensions".

To resume: although there are conceptual differences between knowledge representation and data modeling, there seem to be enough similarities for possibly enabling the use of database techniques for the support of knowledge-based systems.

Why are database functions of interest for knowledge based systems ?

A database system should work like a public library, which keeps books independent of the readers and which provides a way to share books among an arbitrary number of people. Of course, we expect, that the quality of the books is assured, that the collection is up to date and that there is a simple and clean interface for making use of the library.

In more technical terms: The main purpose of a database system is the management of shared data independent of any application program. This includes the following functions:

- Persistent storage.
- Quality assurance, i.e. fault tolerance, concurrency control, authorization, and integrity control.
- Efficient processing even of large quantities, achieved by means of buffering, access paths, set oriented operations, pipelining of operations, etc.
- Implementation independent interface, achieved by means of a high level data model (rules for representing the data at a conceptual level and operations for manipulating this representation).

Of course, persistency can also be provided by a file system, however, file systems do not support data modeling, integrity control, access-paths, and other facilities of a typical database systems.

Since knowledge-engineering, i.e. the building and maintenance of a knowledge base is quite an expensive task, there are enough economic reasons to keep knowledge bases in the same way as libraries and data bases. So, it is clear that database systems provide services needed by knowledge based systems. However, current database technology has been designed with more primitive concepts in mind ("data instead of knowledge").

How can database functions be made available for knowledge based systems?

A straightforward approach is *loose coupling*, i.e. to use an existent database system and to install the knowledge base system on top of it. However, the well-known *impedance mismatch* would lead to such enormous inefficiencies that this solutions is known as being unacceptable.

The next possible step is *tight coupling*, i.e. to build a special purpose backend database system that provides services specifically tuned for the needs of a knowledge based system. This approach reflects the possibilities of the current state of the art; and it has been followed in the design of LILOG-DB, which is discussed in some detail in the main section of this paper.

Tight coupling is not an optimal solution. The users of the specialized backend database system are inference machines which typically switch dynamically between a variety of inference strategies. Since this switching cannot be determined in advance, the database system will have problems in always providing the optimal service. This problem gives raise to the final step of *integration*, where knowledge based system and database management system are integrated into a so-called knowledge base management system. This allows to dynamically adjust the low-level behaviour to the actual needs of the inference mechanisms. However, this final step is still science fiction and requires a lot of additional research.

The rest of this paper is structured as follows. In the next section, we work out what extensions of database modeling capabilities are required for a tightly coupled system like LILOG-DB, which is dedicated to the support of knowledge-based systems, especially those working in natural language processing environments.

This gives us criteria for the design of FLL. The basic concepts of the language are presented in section 3, with a special focus on the modeling capabilities offered by the introduction of open structures and the support of taxonomies by a powerful sort-concept.

In section 4, we give some notes on theory and implementation of *FLL*. After a discussion of related work in section 5, some concluding remarks are given in the final section.

2 Requirements for Advanced Data Modeling

It is widely accepted that the state-of-the-art, as far as practically available systems are concerned, is given by relational database systems([11]), which are theoretically founded in first-order logic. So, what are the concrete problems which inhibit the direct use of relational technology for knowledge based systems?

1.) The most important problem is the restriction that relational systems impose on the *format* of their data: their so-called *first-normal-form* enforces all data to be normalized to flat tables, i.e. no attribute can have a complex term as its value.

This necessity to decompose relations with inherently complex attributes into flat tables is not only cognitively inadequate, the overhead in space resulting from the decomposition is exponential in the worst case.

Thus the deductive database language *FLL* presented in this paper is based on a *complex-object data-model*, the *Feature-Term Data-Model (FTDM)*.

2.) Another restriction that stems from the "filing card paradigm" of traditional database systems is that they enforce the user to completely specify the *schema* of the database, i.e. in the design phase of the database (before any tuple can be inserted), the user has to specify exactly which relations contain which attributes.

Once this schema is given, no data can be entered into the database that do not satisfy the prescribed format. Especially, no additional attributes can be stored.

As we will show by example later on, this is too restrictive e.g. in a natural language processing environment: the descriptions of objects that occur in a text can contain items that simply cannot be completely known before, and thus the terms encoding these descriptions cannot be prescribed by a normal form forbidding the storage of additional information.

Therefore *FLL* relies on a concept of *open structures* which allows a simple and proper treatment of missing as well as of additional information.

3.) If we apply databases to mini-worlds that can be encoded on filing-cards, we can suppose a "flat" structure of the world: all concepts are equally ranked, only limited means of classification are needed.

In contrast, a system supporting high-level discourse about a structured mini-world has to incorporate knowledge about that structure to adequately answer queries.

We can assume that any known object of the mini-world of interest has a representation in the database. However, if we administrated a table for each object-class of the real world (e.g. restaurants, pubs, bars, grill-rooms, drive-in-restaurants, pizzerias...), we would have to keep too many very small tables and to store the same object in multiple tables.

Since too many tables and too much redundancy in the tables are a severe source of inefficiency, such a behaviour has to be avoided. Therefore, our database system has to provide effective support of taxonomies. Thus *FLL* is based on a *sort-concept with multiple inheritance*.

4.) The field of logic based database languages has been opened by the work on *DATALOG* ([38]). Starting with *DATALOG*, *FLL* can be described by the following schema:

$$
\begin{array}{ll}
& DATALOG \\
+ & \text{complex objects} \\
+ & \text{some built-ins for arithmetics and comparison} \\
+ & \text{negation and set-grouping} \\
+ & \text{open structures} \\
+ & \text{order-sorting} \\
\hline
= & FLL
\end{array}
$$

The semantics of *FLL* are set-oriented, i.e. all answers to a query are retrieved and presented at a time, in contrast to (e.g.) Prolog, where the results are computed term-at-a-time via backtracking. *FLL*'s kind of negation is stratified negation (see [32]).

The concepts of open structures and of filtering data via order-sorting are the main extensions of *FLL* w.r.t *DATALOG* as well as the language LDL ([6]) having received much attention in the recent years.

They will be the focus of our presentation, which is an informal motivation of the basic new concepts of *FLL*. For a formal and more complete justification, the interested reader is referred to [21].

3 Basic Concepts of FLL

3.1 Complex Objects

In *DATALOG* we cannot express the structure of a stored object except by flattening it and storing the components into auxiliary base predicates. For a number of reasons it seems better to use function symbols as non-atomic arguments and e.g. write

> $father(bob, [peter, paul, mary])$
> ...
>
> $siblings(X) : -father(Y, X)$

instead of

> $father(bob, peter)$
> $father(bob, paul)$
> $father(bob, mary)$
> ...
>
> $siblings(X) : -father(Y, X)$

especially when all *siblings* are accessed simultaneously.

Besides reasons of compact storage and fast access of complex objects as a whole, for reasons of cognitive adequacy it seems convenient not to force a user to artificially decompose objects which (s)he sees as a whole.

Since *FLL* will have an interface to our Prolog-implementation *TLPROLOG* (see [24]), it has to support at least **lists** and **functors**, and that are the first complex objects we introduce.

We do not discuss them in detail here because they by now are a standard of deductive database languages. This holds for **sets** too. If we don't want to represent the order of children of *bob* in a list, in *FLL* we can write

> $father(bob, \{peter, paul, mary\})$

instead of the above fact as well.

3.2 Negation and Set-Grouping

The most important aspects of *FLL* shall now be briefly overviewed, mainly in comparison with *DATALOG*.

Recursion: The need for recursion in first-order databases has often been stressed in the literature (see [4,6]). E.g. we may want to compute the set of all cities reachable from each other, using the predicate

```
reachable(X, Y) :- way(X, Y).
reachable(X, Y) :- way(X, Z), reachable(Z, Y).
```

The answer to

```
?- reachable(X, Y).
```

then may be the table

X	Y
a	b
a	c
a	d
b	c
b	d
c	d

In *FLL*, as in *DATALOG*, we have the possibility to write arbitrary recursive programs, i.e. there are no restrictions to linear recursion or whatever else.

Set-Grouping: What can we do if we want a list like

X	Y
a	b,c,d
b	c,d
c	d

instead of the answer above? Till now our language doesn't enable us to get such an answer since we are unable to explicitly (dis-)aggregate sets. *DATALOG* does not distinguish different levels of *set-grouping* [6,36]. Thus we augment *FLL* by a set-grouping operator and state the query

```
?- reachable(X, < Y >).
```

which means that our answers are to be grouped by Y. Our result is the intended table shown above.

In contrast to the lists and functors introduced above, sets have no internal structure. They are the adequate representation for large collections of entities which do not fit into the more or less tree-like structure of lists and functors.

Built-In Predicates: In the context of sets, *FLL*'s built-in predicate *in* is of special importance. As the rule

```
child_of(X, Y) :-
    father(Y, CHILDREN),
    X in CHILDREN.
```

illustrates, it performs an un-grouping of sets.

Other built-in predicates include arithmetics and comparison, as illustrated by the self-explaining rules

```
sum(X, Y, Z)    :-    Z is X + Y.
max(X, Y, X)    :-    X > Y.
max(X, Y, Y)    :-    X ≤ Y.
```

Negation: Another aspect of the missing expressive power of *DATALOG* for our purposes of knowledge-processing is that we cannot handle negation. We are not able to state predicates like

```
single(X) :- person(X), not(married(X)).
```

Thus we introduce *stratified negation* [6,32,34] into *FLL*. That means that we allow predicates like the one defined above with negated literals on the right-hand side of a rule if this negation is not part of a recursive cycle as in the following case:

```
even(0).
even(I) :- not(odd(I)).
odd(succ(I)) :- even(I).
```

This is because such an unstratified program might not have a unique minimal model, i.e. it's got no proper declarative semantics (this is oversimplified, see [32] for a discussion).

3.3 Feature-Tuples, and the Open World Assumption

From the point of view of a database programmer, tuples are a very convenient data-structure for expressing roles of objects. This is because for tuples, in contrast to previously introduced structures like lists or functors, these roles can be adressed by name.

It appears more natural to model the information on a book by a tuple

$\langle$ *author* : '*Hammett*', *title* : '*RedHarvest*' $\rangle$

and refer to the author via the attribute name *author* than to represent it as a functor

book('*Hammett*', '*RedHarvest*')

and adress the author by the position number 1.

Additionally, tuples match the intuition of a table in which the human user often imagines knowledge.

So, it is a natural idea to extend *DATALOG* by tuple-aggregation to achieve greater ease and intuitiveness of database programming. E.g. a program like

> *bookinfo*($\langle$ *author* : X, *title* : Y, *isbn* : Z $\rangle$) :-
> *writer*($\langle$ *author* : X, *title* : Y $\rangle$),
> *book*($\langle$ *title* : Y, *isbn* : Z $\rangle$).

would be much simpler to read and maintain then

> *bookinfo*(X, Y, Z) :-
> *writer*(X, Y),
> *book*(Y, Z).

since the user would not be forced to remember the semantics of argument positions. This argument is of great practical importance when one writes large scale applications on a logic database. Similar arguments for the introduction of tuple-like structures into logic-based languages can be found in [3,7].

So far, we have freed the programmer from the need to express roles via position numbers. But if we interpret tuples a usual, (s)he still has to specify irrelevant information. E.g. if the schema of the base-relation expressed by *book* were

$\langle$ *author* : .., *title* : .., *subject* : .., *keywords* : .. $\rangle$

a program would have to look like[1]

> *bookinfo*($\langle$ *author* : X, *title* : Y, *isbn* : Z $\rangle$) :-
> *writer*($\langle$ *author* : X, *title* : Y, *subject* : _, *keywords* : _ $\rangle$),
> *book*($\langle$ *title* : Y, *isbn* : Z $\rangle$).

to obtain the required information, although *subject* and *keyword* are completely irrelevant for *bookinfo*.

The reason for this is that ordinary tuples are interpreted as *closed structures*:

> a tuple can *only* be matched by a tuple that contains exactly the same attributes, if all these values match

To overcome this deficiency, in *FLL* we introduce the concept of **feature-tuples**. Their name stems from the origin of feature-tuples in the *feature-structures* of knowledge representation and the tuples of Relational Algebra:

> A feature-tuple t_1 can be matched by any feature-tuple t_2 if all their common attributes (called features) can be matched.

With feature-tuples, we can write

> *bookinfo*($\langle$ *author* : X, *title* : Y, *isbn* : Z $\rangle$) :-
> *writer*($\langle$ *author* : X, *title* : Y $\rangle$),
> *book*($\langle$ *title* : Y, *isbn* : Z $\rangle$).

simply ignoring the irrelevant features *subject* and *keyword*.

Using variables ranging over feature-tuples, we can now very conveniently express the *collection of information in open structures*. A feature-tuple variable X in the head of a rule is instantiated to a tuple collecting all features of all instances of X in the rule-body. E.g. if we have

> *writer*($\langle$ *authorname* : *agatha_christie*,
> *nationality* : *british* $\rangle$).

and

---

[1] _ stands for the anonymous variable

$$book(\ \langle\ \ title:\ 'TheRedKimono',$$
$$authorname:\ agatha\_christie\ \ \rangle\ ).\ \ ,$$

for the rule

$$bookinfo(W):\text{-}\ writer(W),\ book(W).$$

and the query

$$?\text{-}\ bookinfo(X).$$

we obtain

$$X\ =\ \langle\ \ \ title:\ 'TheRedKimono',$$
$$authorname:\ agatha\_christie,$$
$$nationality:\ british\rangle$$

The example shows that by means of feature-tuples, we have introduced a *natural join* into deductive databases, which permits the collection of information in open structures.

It allows a convenient description of the refinement of information (e.g. *bookinfo*) from different sources (*writer*, *book*).

So, the basic paradigm of computation for feature-tuples is compatibility of information, instead of equality of information for closed tuples: the processing of tuples continues until their information grows inconsistent, i.e. contradicting values for a feature are detected.

Applied to the treatment of feature-tuples in knowledge-bases, our sight is an open-world assumption (*OWA*) on *knowledge-elements* since feature-tuples do not specify all information about an object, but only that which is known (to the system). This information can be augmented as long as the result stays consistent.

Note that the OWA on knowledge-elements has to be distinguished from the OWA *on knowledge-bases* themselves. The first one says that our knowledge-elements are incomplete descriptions of entities of the world; the latter one states that the knowledge-base itself is incomplete, i.e. does not contain all propositions that are true for the entities.

We claim that the reliance on the more flexible notion of compatibility instead of equality allows a simpler, more convenient and more natural description of the deduction of knowledge from a database.

As the later discussion will show, this argument grows even stronger if these open structures are augmented by a sort-concept describing which information is adequate and/or required in which context.

3.4   Sorts

The necessity to incorporate taxonomic information in next-generation DBMS has often been stressed in the literature (a survey on taxonomies in semantic data-models can be found in [31]).

Expressing taxonomies is a requirement of adequate modeling, since most relevant mini-worlds are not flat-structured, but divided into concepts and sub-concepts.

E.g. the field of *logic database languages* is a part of *query languages*, which is a discipline of *databases*, which itself is a subdiscipline of *computer science*.

In *FLL*, we can model this as follows:

$$sort\ logic\_database\_languages\quad isa\ query\_languages$$
$$sort\ query\_languages\qquad\qquad\ isa\ databases$$
$$sort\ databases\qquad\qquad\qquad\ isa\ computer\_science$$

A crucial question when incorporating support of taxonomies in a data-model is how taxonomies are interfaced to the rest of the program. In *FLL*, we introduce the .-operator. In a program, $X.s$ denotes that all terms passed through X have to be of sort s.

E.g. suppose the taxonomy described above, we can write the rule

$$database\_researcher(\ \langle name:\ X\ \rangle\ ):\text{-}$$
$$researcher(\ \langle name:\ X, topic:\ Y.databases\ \rangle)$$

filtering out the names of all researchers that work in database systems.

However, the above rule cannot satisfy us. The reason is that we have not yet specified how terms of the sort *database* look like, so that the filtering-operation $Y.databases$ is rather useless.

Since we want to use our sort-concept for purposes like strong typing and the provision of conceptual schemas for sets, we need a means of expressing the structure assumed for the terms belonging to a sort.

Sort Descriptions

So, we introduce *sort-descriptions* for the structure of terms belonging to a sort.

In the following example, we represent some coloured geometric objects: circles are described by colour and radius, rectangles by the lengths of their sides, and colours are one of *red, blue* etc.

$$
\begin{array}{lll}
sort\ colours & = & \{red, blue, yellow, green\} \\
& & isa\ atom \\
sort\ circle & = & tuple\ (radius : real \\
& & \qquad colour : colours) \\
& & isa\ tuple \\
sort\ rectangle & = & tuple\ (side1 : real \\
& & \qquad side2 : real \\
& & \qquad colour : colours) \\
& & isa\ tuple
\end{array}
$$

The example illustrates that sorts can be described by enumerations of terms or by structural descriptions. E.g. the members of the sort *rectangle* are tuples whose features *side1, side2* are restricted to the (built-in) sort *real*[2], and *colour* has to be one of *colours*.

Supposing we got a predicate *object* representing a set of geometric objects, we can retrieve all circles which fit into a rectangle of the same colour by

$$
\begin{array}{l}
inner\_circle(X) :\text{-} \\
\quad object(\ (side1 : S_1, side2 : S_2, colour : C).rectangle\ ), \\
\quad object(\ X.circle\ ), \\
\quad X = (radius : R, colour : C\ ), \\
\quad S_1 \geq R, S_2 \geq R.
\end{array}
$$

The example shows how sort-restrictions act as filters: only those *objects* are passed to the checks at the end of the rule which satisfy the sort-restrictions of *rectangles* resp. *circles*. But note that the order of the subgoals of a rule doesn't determine the semantics. In the example, the subgoals have been written in the order which the FLL-Compiler would choose for the evaluation of the rule-body to obtain an efficient *sideways information passing* between the variables of the rule (see [41]).

Constraints

A full sort-specification of the *FTDM* is of the form

$$
\begin{array}{ll}
sort\ \text{SORTNAME} & = \quad \text{SORTDESCRIPTION} \\
& with\ \text{CONSTRAINT} \\
& isa\ \text{SORTNAMELIST};
\end{array}
$$

In the CONSTRAINT-part, semantic constraints can be imposed on the terms of the sort. We can e.g. describe squares by

$$
\begin{array}{lll}
sort\ square & = & tuple\ (side1 : real \\
& & \qquad side2 : real \\
& & \qquad colour : colours) \\
& with & side1 = side2 \\
& isa & rectangle
\end{array}
$$

Constraints allow us to incorporate semantic knowledge into the description of sorts. Some more examples will be given in the discussion.

Sort Lattices and Multiple Inheritance

In the above example, the sort *square* is a specialization of the sort *rectangle*. Nevertheless, in its description we repeated the whole description of *rectangle*, only specializing it by the constraint *side1 = side2*.

---

[2]The predefined, built-in sorts and their denotations are

term	the whole *FTDM*-universe
nil	the empty set
integer	the set of integers
real	the set of real numbers
atom	the set of atoms, i.e. names

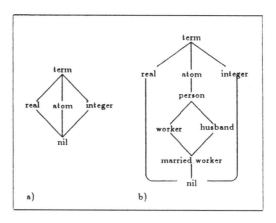

Figure 1: The Default Sort-Lattice and the Embedding of User-Defined Sorts

This inconvenience is unnecessary, since a sort inherits all properties of its supersorts. So, without loss of information we could have written

>*sort square*   *with*   *side1 = side2*
>   *isa*   *rectangle*

simply omitting the repeated sort-description. Note that a sort can have multiple supersorts, so that we obtain *multiple inheritance.*

To properly organize this multiple inheritance mechanism, we require the sort-specifications of a program to form a lattice with respect to the *isa*-relation.

The lattice structure basically requires that for each two sorts, their smallest common supersort and their greatest common subsort are uniquely determined.[3] That e.g. excludes cyclic *isa*-relations and eases the detection of incorrect typing, i.e. the existence of sorts with no members.

An example of such a sort-lattice, spanned by the declarations

>*sort person* $=$ *tuple (name : atom, age : integer)*
>   *with age > 0 and age < 150*
>   *isa atom .*
>*sort worker* $=$ *tuple (job : atom, salary : integer)*
>   *isa person .*
>*sort husband* $=$ *tuple (partner : husband)*
>   *isa person .*
>*sort married_worker*
>   *isa husband, worker*

is depicted in Fig.1b.

Note that for *married_worker*, neither a sort-description nor a constraint is given. An omitted sort-description stands for "$= term$", an omitted constraint for "*with true*", and omitted isa-specifications for "*isa term*".

Fig.1a shows the *default lattice* of our *Feature Term Data Model*, i.e. the structure of predefined sorts. The declarations of user-defined sorts are inserted into the default-lattice so that *term* always remains the top sort and *nil* is an immediate subsort of any sort for which no subsort is specified.

3.5   Discussion

We will now discuss the most important aspects of the use of sorts and feature-tuples in *FLL*, focusing on the interaction of these two concepts.

---

[3]In fact, for the purposes of *FLL* it would be sufficient to require the sorts to form a partial order. The lattice requirement stems from the use of sorts in order-sorted unification in the logic levels of of *LILOG-DB*, on top of the database kernel.

Conceptual Modeling: Let us reconsider the rule

$single(X) :- person(X), not(married(X))$.

we used for the introduction of negation. We could as well have modeled *persons* and *singles* by the sort-specifications[4]

```
sort person = tuple(   name : atom,
                       firstname : atom,
                       husband : person or { none },
                       children : setof person,
                       sex : { male, female },
                       age : integer )
       with age > 0 and age < 150 and
         □ has name and □ has firstname and
         ( forall V in children : V.name = name)
       isa term


  sort single
       with husband = none
       isa person
```

This example requires some discussion on the use of sorts for conceptual modeling:

1.) The example exhibits some more of the modeling cababilities of our sort-concept.

1. Note that *person* is a recursively defined sort since the feature-specifications for *husband* and *children* refer to other *persons*.

2. In the declaration of the feature *children*, the sort-constructor *setof* has been used: *children : setof person* denotes that the value of the feature *children* has to be a (possibly empty) set of *persons*.

3. The specification { *male, female* } for the *sex* of *persons* restricts the admissible values of the feature *sex* to one of the constants *male* or *female*. It shows that finite sets of terms can serve as sort-constants.

4. The constraint for *person* is a boolean conjunction of five simple constraints. Not only conjunction, but boolean negation and disjunction too are allowed for the combination of constraints and sort-descriptions. Another example is the limitation of *husbands* to either *persons* or the constant *none*.

5. The constraint □ *has name* and □ *has firstname* enforces that all terms of sort *person* have to have at least the features *name* and *firstname* specified. This restricts openness of feature-tuples by specifying which information is at least required for proper processing. E.g.
⟨ *firstname : jim, age : 35* ⟩ is not a member of this version of sort *person* since no *name* is specified.

6. In contrast, the term
⟨ *name : jones, firstname : jim, job : waiter* ⟩ is a member of the sort *person* since the sort-specification specifies *persons* as (open!) feature-tuples, and no restriction on a possible feature *job* is imposed. This shows that sorts allow for the description of incomplete as well as of *additional* information.

7. Finally, observe that we have made use of universal quantification by stating that all *children* of a *person* have to have the same *name* as the *person* itself.

So, our little example exhibits most of the relevant means for the description of sorts in *FLL*.

2.) Sorts are a concept for describing *properties* of terms. Unlike most object-oriented systems, *LILOG-DB* does not administrate extensions of sorts. Sorts cannot be queried, but only be used as filters during query-processing.

In general, sorts express the conceptual structure of a mini-world, while rules work on the concrete database, i.e. the stored extensions.

Sorts free the user from expressing the taxonomic structure of a mini-world by hand, and leave only the "hard" deduction tasks – the derivation of new knowledge from existing one – to rules.

---

[4] □ stands for the root term considered

In the above example, encoding of the *single*-property in a rule requires an unnecessary join which is inefficient and inadequate since the *single*-property is a specialization of the *person*-property that can be checked by filtering.

So, replacing filtering with predicates by filtering with sort-restrictions can significantly speed up query-processing: unnecessary joins are replaced by on-the-fly filtering.

Incomplete and Variant Schemas: In the *FLL*-environment, sets stored in the database are called *ft-sets*. They are created by statements like

$$create\_ft\_set(\langle name : ftsetname, schema : sortname \rangle)$$

That means that sorts serve as conceptual schemas of *ft-sets*.

The facilities for the declaration of sorts include the possibility to express incomplete or variant sort-information. Incompleteness is used in the specification of feature-tuple sorts. E.g. the sort-declaration

$$sort\ person \quad = \quad tuple\ (\quad name : atom,$$
$$age : integer\ )$$
$$isa\ term$$

is incomplete since the *OWA* allows terms of the sort *person* to have more than the two features *name* and *age*, e.g. the term $\langle name : john, age : 32, job : waiter \rangle$ is of the sort *person*. In contrast, a "closed" sort *person* would look like this:

$$sort\ person \quad = \quad tuple\ (\quad name : atom,$$
$$age : integer\ )$$
$$with \qquad \Box\ hasarity\ 2$$
$$and\ \Box\ has\ name$$
$$and\ \Box\ has\ age$$
$$isa\ term$$

In *FLL*, this can be abbreviated to

$$sort\ person = closed\ tuple\ (name : atom,\ age : integer\ )$$
$$isa\ term.$$

Sort-declarations may be variant too, i.e. disjunction can be expressed. The sort

$$sort\ person \quad = \quad tuple\ (\quad name : atom,$$
$$age : integer\ or\ atom\ )$$
$$isa\ term$$

allows the expression of *age* by a number or a textstring.

A more convincing example returning to the objects of geometry is the following: it is well known that triangles are completely described by two sides and an angle or by two angles and a side. This can be captured by a variant sort-specification looking like this:

$$sort\ triangle = tuple\ \langle$$
$$side\_a : real,$$
$$side\_b : real,$$
$$angle\_ab : degree\rangle$$
$$or \quad tuple\ \langle$$
$$side\_a : real,$$
$$angle\_ab : degree,$$
$$angle\_ac : degree\rangle$$
$$with\ \Box\ hasarity\ 3$$

As far as schemas are concerned, the introduction of feature-tuples as open structures is the most crucial point in our argumentation: *Wouldn't optional fields in tuples be sufficient?* The answer, especially in the context of natural language based knowledge-processing, is: no.

Suppose we want to store all cars referred to in a text in an ft-set *cars*. We may introduce a sort

$$sort\ car = tuple\ (type : atom,\ speed : integer\ ..)$$

and then create an ft-set in the database by

$$create\_ft\_set(\langle name : cars, schema : car\rangle).$$

But what do we do with a piece of text like:

Jim has a red car with a dent in the right fender.

We argue that in a general natural language processing context, one simply cannot predict all possible characterizations of *cars* and describe them as optional fields in a database schema.

The consequence of this argument for the processing of the database-schemas in *LILOG-DB* is that our storage system has to cope with a much smaller degree of meta-knowledge about data than existing DBMS can exploit. Storage techniques of *LILOG-DB* are discussed in [19].

Strong Typing, and more: By the sort-concept, *FLL* is strongly typed. Sort-restrictions on ground terms (e.g. "1.*integer*") are statically checked by the *FLL*-compiler, and the use of sorted variables invokes some elaborate rule-processing.

E.g. suppose the sort-specifications

$$sort\ european\quad =\quad \{\ british, german\ \dots\ \}.$$
$$sort\ asian\quad =\quad \{\ chinese, indian\ \dots\ \}.$$
$$sort\ eurasian\quad\quad isa\ european, asian.$$
$$sort\ american\quad =\quad \{\ us, mexican\ \dots\ \}.$$

Then a rule like

$$transatlantic\_country(X) :\text{-}$$
$$country(X.european),$$
$$country(X.american).$$

is rejected since it is detected inconsistent: the sorts *american* and *european* – to which the same variable X is restricted – are disjoint.

Sort-restrictions are used for optimization too: For the rule

$$p(X.eurasian) :\text{-} q(X.european), r(X).$$

we know that $eurasian < european$ and can replace $q(X.european)$ by $q(X.eurasian)$ and push the sort-restriction into the body of the rule:

$$p(X.eurasian) :\text{-} q(X.eurasian), r(X.eurasian).$$

Both the rejection of rules inconsistent with the sort-declarations and the selection of the strongest possible sort-restrictions avoid the incorporation of irrelevant terms in intermediate results, thereby speeding up query-processing.

4 Some Notes on Theory and Implementation of FLL

4.1 Theory

The theory of *FLL* has been developed in [21], were straightforward denotational semantics for sorts and minimal model semantics for order-sorted programs are given. It has been previously stated that sorts in *FLL* are used as filters. In this context, it is of special importance that for a term t it is *decidable in linear time* whether t is contained in a sort or not.

In [21], it has been shown that the sort-concept somehow lies *under* the bottom-up semantics of Horn-Clause programs: only the notion of a substitution has to be replaced by that of a *well-sorted substitution for the FTDM*, while the usual minimal model semantics remain unchanged.

Finally, for stratified programs, i.e. programs which use neither negation nor set-grouping inside of recursive cliques, it can be shown that there always exists a *minimal model*. It can additionally be shown that perfect-model semantics ([32]) can be applied to *FLL*-programs in a straightforward manner.

A note on expressive power: In *FLL*, one can write a program which counts the cardinality of a set. This – in connection with the power of recursion and the fact that feature-tuples are unranked – allows to show that *FLL* is of Turing complexity, i.e. anything that is computable can be programmed in *FLL*.

4.2 Implementation

FLL is a part of the *LILOG-DB* system, which is implemented in C. The implementation of *LILOG-DB* consists of five components: The *Logic Programming System* (LPS) provides the interfaces of *LILOG-DB* to top-level inference-engines. The *Sort Lattice Manager* (SLM) and the *Rule System Manager* (RSM) are modules for the specially tuned administration of large sort-lattices resp. rule-systems. The *Query Processor*

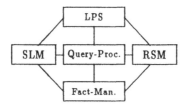

Figure 2: Architecture of LILOG-DB

provides means for efficient processing of complex database queries, using the *Fact Manager* to administrate sets stored on disk.

FLL is part of the *Logic Programming System*. *FLL* programs are processed in three phases.

1. first, they are globally optimized with respect to certain query forms which shall be most effectively supported.

2. for a given query, an operator tree of our database algebra *EFTA* ([22,27]) is generated by the *FLL* -Compiler ([26]).

3. This operator tree is submitted to the *Query Processor*, which performs local optimization and generates a *Query Evaluation Plan*, which is then evaluated against the database. The access to *ft-sets* stored on disk is performed by the *Fact Manager*.

During this process, sort-information is exploited at any stage:

- in the compilation-phase, programs are statically type-checked

- sorts serve as conceptual schemas for ft-sets; they are used to obtain compact internal schemas and optimize access

- sortal information is further exploited to optimize the representation of intermediate results in query processing and to push selective filter-operations as deep as possible

Details on principles and techniques of the implementation can be found in [19,23,28,26,25].

5 Related Work

Recent work in deductive databse languages shows a growing tendency to incorporate results of many other disciplines of computer science, such as logic, programming languages, AI etc. This yields very fruitful new approaches to database systems and database programming, but makes it sometimes hard to account for even the most important relations between each two systems.

So in the following we will not try to compare *FLL* with some specific languages, but isolate some of the main streams of research and locate the place of *FLL*, without claiming completeness. We make six points, which are visualized in Fig. 3.

1.) The sort-concept of *FLL* can be very well seen as a semantic data-modeling capability. Semantic Data Models (see [31]) arose as an approach to incorporating more semantics into the conceptual schemas of relational database systems.

They offer facilities like direct support for relationships (see [10]), inheritance, constraints or data-abstraction, which are not captured by the relational model.

This yields a more direct and convenient way from the mental image of an application to its encoding in a database program, and queries can be stated in a more compact and natural way.

However, the implemented systems differ very much in the aspects of data-modeling they support. E.g. SDM ([16]) offers class-abstraction using a powerful apparatus of modeling capabilities, but does not represent relationships between classes. In contrast, TAXIS ([30]) focuses on specialization-hierarchies and classification.

The sort-concept of *FLL* tries both: to provide sufficient modeling capabilities for data-abstraction and adequate support of hierarchies.

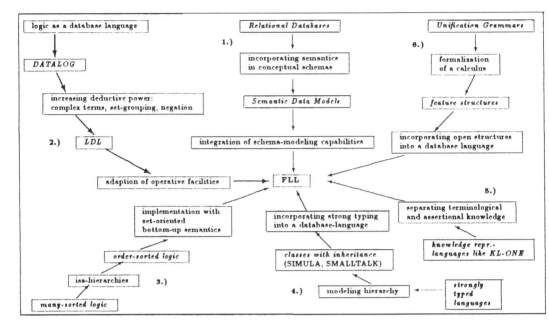

Figure 3: *FLL* in the Context of some Related Work

Nevertheless, one aspect of semantic data modeling is still missing: the current version of *FLL* lacks a concept of *integrity constraints*[5] which adequately capture the relationships between *ft-sets*. A prototype concept has been worked out, but is not yet implemented.

2.) The field of *logic as a database language* has emerged in the late seventies. The most prominent language resulting from this approach is *DATALOG* ([38]), which basically is function-free Horn-Clause logic, interpreted with set-oriented bottom-up semantics.

For the requirements of a full-scale deductive database system, this is not expressive enough. In the language LDL ([6]) having received much attention in the database world, the expressive power has been increased by augmenting *DATALOG* by complex terms, set-grouping and negation.

FLL has adapted these operative facilities, extending them by open structures (feature-tuples) and an underlying sort-concept.

3.) This leads us into the field of order-sorted logic. One can say that what LDL is for unsorted logic is *FLL* for order-sorted logic: an implementation with set-oriented semantics and extended operative facilities like set-grouping or negation.

4.) Order-sorted logic itself is an extension of many-sorted logic where the sorts are ordered by an inheritance hierarchy.

Similar approaches are well-known in the world of programming-languages, where the possibility of modeling hierarchy in strongly typed languages has been incorporated in languages allowing for classes with inheritance, the earliest one being SIMULA ([12]). From a programming language point-of-view, the sort-concept of *FLL* incorporates strong typing in a logic database language.

The same has been done in many object oriented database programming languages, such as O++ or E (based on C++, see [2,33]).

5.) The separation of types ("sorts", "classes") and values is a feature that can be found in knowledge-representation languages such as KL-ONE ([8]). In KL-ONE, assertional knowledge ("rules") is stored in a so-called A-Box while terminological knowledge (information on the structure of the mini-world, "sorts") are modeled in a T-Box.

---

[5]Note that *local* integrity constraints on single *ft-sets* can be expressed by sort-constraints; see the *person*-example above.

The strict separation between structural background knowledge and rules working on that structure is a very popular principle in knowledge representation which has been adopted for *FLL*, which – as previously mentioned – is designed for database programming in a knowledge representation environment.

6.) A last input stream of our work shall be acknowledged. Open structures, so-called feature-structures, originated when the operational capabilities of the unification grammars ([35]) of computational linguistics were formalized to a calculus (see [18]).

The feature-tuples of *FLL* are an attempt to incorporate such open structures into a database language.

We hope to have shown that these open structures provide valuable means for convenient, natural and efficient knowledge representation in *FLL*.

6 Conclusion

In this paper, we have discussed the modeling capabilities required for the support of knowledge-based systems by next-generation database systems.

Especially, we have motivated and discussed the basic concepts of *FLL*, a deductive database language implemented as a part of our project *LILOG-DB*.

FLL is based on the *FTDM*, a complex-object data-model supporting open structures and the processing of taxonomies.

It has purely declarative set-oriented bottom-up semantics, with the operational capabilities of recursion, stratified negation and set-grouping.

Open structures are introduced via the concept of feature-tuples, allowing adequate modeling and efficient processing of incomplete information.

The basic paradigm behind the introduction of feature-tuples is that information can be combined iff it is *compatible*. This is far more flexible, convenient and natural than the reliance on the notion of *equality* that can be found in most *DATALOG*-successors.

The sort-concept of *FLL* allows for the representation of the logical structure of a mini-world.

By *sort-descriptions* and *constraints*, structural and semantic knowledge on the entities of a world can be expressed in a uniform and convenient way.

The multiple inheritance mechanism on sort-lattices allows a proper and natural description of taxonomic hierarchies.

Sorts in *LILOG-DB* serve as conceptual schemas of sets stored on disk as well as of intermediate results and enable the derivation of compact internal representation schemas for sets.

Additionally, they type *FLL*-programs, so that the full corrective power of strong typing is made available to logic-based database programming.

Finally, our notion of sorts covers a lot of "local" integrity constraints.

In connection with feature-tuples, our sort-concept allows for the description of *incomplete and variant schemas*, which is an inevitable necessity in a knowledge representation and natural language processing environment.

The implementation of *FLL* is part of the *LILOG-DB*-system, which has been implemented in *C. FLL* is compiled to *EFTA*, the algebra of the *LILOG-DB* query processor. Details of the implementation can be found in [26].

Sources of inspiration for the basic concepts of *FLL* have been made out in order-sorted logic, the schema modeling capabilities of semantic data-models, the open structures of natural language processing, the operational facilities of *DATALOG* and *LDL*, and complex object extensions of relational algebra. This related work has been discussed in detail.

References

[1] J. R. Abrial, "Data Semantics", in J. W. Klimbie, K. L. Koffeman (eds.): *Data Base Management* North Holland, 1974

[2] R. Agrawal, N. Gehani: *ODE (Object Database and Environment): The Language and the Data-Model* SIGMOD 1989

[3] H. Ait-Kaci, R.Nasr: LOGIN: A Logic Programming Language with Built-In Inheritance JLP 1986, 3, 185-215

[4] F. Bancilhon, R. Ramakrishnan: An Amateur's Introduction to Recursive Query Processing Strategies SIGMOD, Washington, 1986, pp. 16-52

[5] C. Beeri: "Formal Methods for Object Oriented Databases"Proc. 1st Int. Conf. on Deductive and Object-Oriented Databases, 1989

[6] C. Beeri et al.:   Sets and Negation in a Logic Database Language (LDL1) PODS 6, 1987, 21-37

[7] C. Beeri, R. Nasr, Sh. Tsur: Embedding ψ-terms in a Horn-Clause Logic Language Proc. 3rd Int. Conf. on Data and Knowledge Bases, Jerusalem, 1988

[8] R. Brachman, J. Schmolze: An Overview of the KL-ONE Knowledge Representation System. Cognitive Science 1985, 9, 171-216

[9] S. Ceri, G. Gottlob, L. Tanca: "Logic Programming and Databases" Springer-Verlag, 1990

[10] P. Chen: The Entity-Relationship Model: Toward a Unified View of Data. TODS 1976, Vol. 1, No. 1, 9-36

[11] E.F. Codd: A Relational Model for Large Shared Data Banks CACM 13:6, pp. 377-387

[12] O.J. Dahl et al.:   SIMULA 67 Common Base Language Publication N.S-22, Norwegian Computing Center, Oslo, 1970

[13] R. Elmasri, S. B. Navathe: Fundamentals of Database Systems Benjamin/Cummings, 1989

[14] R. A. Frost: Introduction to Knowledge Base Systems Collins, 1986

[15] G. Gardarin, P. Valduriez: Relational Databases and Knowledge Bases Addison-Wesley, 1989

[16] M. Hammer, D. McLeod: Database Description with SDM: A Semantic Database Model. ACM TODS 1981, Vol. 6, No. 3, 351-386

[17] H. Kamp: Discourse Representation Theory: What it is and where it ought to go. LNCS, Vol. 320, 84-111

[18] R.T. Kasper, W.C. Rounds: A Logical Semantics for Feature Structures Proc. 24th Ann. Meeting of the Ass. for Comp. Linguistics, New York, 1986, 257-265

[19] M. Ley: Ein Fact-Manager zur persistenten Speicherung variabel strukturierter komplexer Objekte IBM Germany, LILOG-Report 60, 1988, in German

[20] M. Ley, B. Walter: Der LILOG-DB Fact-Manager: Ein Datenbankkern zur Speicherung variabel strukturierter komplexer Objekte to appear in: Informatik - Forschung und Entwicklung, 1990

[21] T. Ludwig: FLL: A First-Order Language for Deductive Retrieval of Feature Terms IBM Germany, LILOG-Report 57, 1988

[22] T. Ludwig: EFTA: An Algebra for Deductive Retrieval of Feature Terms IBM Germany, LILOG-Report 58, 1988

[23] T. Ludwig: Algebraic Optimization of EFTA -Expressions. IBM Germany, LILOG-Report 59, 1988

[24] T. Ludwig: The Design and Implementation of TLPROLOG IBM Germany, IWBS-Report 97, 1989

[25] T. Ludwig: A Brief Overview of LILOG-DB . Proc. Data Engineering Conf. 1990

[26] T. Ludwig: Compilation of Complex Datalog with Stratified Negation Proc. 9th British National Conference on Databases, York, 1990

[27] T. Ludwig, B.Walter: EFTA : An Algebra for Deductive Databases. To appear in: Journal of Data and Knowledge Engineering

[28] T. Ludwig, B.Walter, M.Ley, A.Maier, E.Gehlen: LILOG-DB: Database-Support for Knowledge-Based Systems Proc. BTW, March 1989, Springer

[29] S. Naqvi, S. Tsur: A Logical Language for Data and Knowledge Bases Computer Science Press, 1989

[30] B. Nixon, L. Chung, et al.:   Implementation of a Compiler for a Semantic Data Model: Experience with Taxis. SIGMOD Conf. 1987, San Francisco, 118-131

[31] J. Peckham, F. Maryanski: Semantic Data Models. ACM Computing Surveys, Vol.20, No.3, Sept. 1988

[32] T. Przymusinski: On the Declarative Semantics of Deductive Databases and Logic Programs In: Foundations of Deductive Databases and Logic Programming, Morgan Kaufmann, 1988, pp. 193-216

[33] J.E. Richardson et al.:   The Design of the E Programming Language Comp. Science Technical Report 824, Univ. Wisconsin, Madison, 1989

[34] J.C. Shepherdson: *Negation in Logic Programming* In: Foundations of Deductive Databases and Logic Programming, Ed.: J. Minker, Morgan Kaufmann, 1988, pp. 19-88

[35] S. Shieber: *An Introduction to Unification-Based Approaches to Grammar* CLSI Lecture Notes 4, Stanford University, 1986

[36] O. Shmueli, Sh. Naqvi: *Set Grouping and Layering in Horn-Clause Programs* 4th International Conference on Logic Programming, Melbourne, 1987, 152-177

[37] J. F. Sowa: *Conceptual Structures, Information Processing in Mind and Machine* Addison-Wesley, 1984

[38] J. Ullman: *Principles of Database Systems.* Computer Science Press, 1987

[39] J. D. Ullman: *Principles of Database and Knowledge-Base Systems, Vol. I* Computer Science Press, 1988

[40] J. D. Ullman: *Principles of Database and Knowledge-Base Systems, Vol. II* Computer Science Press, 1989

[41] C. Zaniolo and D. Sacca: *Rule Rewriting Methods for Efficient Implementations of Horn Logic* MCC Technical Report DB-084-87, 1987

Integrity and recursion:
two key issues for deductive databases

Rainer Manthey

European Computer-Industry Research Centre (ECRC)

Arabellastr. 17, D-8000 München 81, Germany

rainer@ecrc.de

1 Introduction

The notion of a *deductive database* has become rather popular during the last decade as a generic name for numerous approaches aiming at an extension of databases with rules. A kind of smallest common denominator of all these approaches is the exploitation of certain forms of general laws for an intensional representation of data. Assume, for example, that a general dependency like "Every employee is supervised by the manager of his department" holds in a company. There is no need to extensionally store facts stating who are the supervisors of each individual employee in the personnel database of that company, as these facts can be derived from the information about assignment of employees to departments, and of departments to managers. General laws of this kind are ubiquituos: they can be found in nearly every domain of application.

The AI community has been studying techniques for handling general knowledge rather intensively since long time. Expert systems, diagnostic systems, and many related paradigms are fundamentally based on some kind of rule notion. The practical relevance of the rule paradigm is meanwhile well accepted in this community. Rule-based AI systems, however, have hardly ever been applied in combination with large sets of facts kept on secondary storage. Expert systems traditionally are main-memory systems efficiently applicable only to problems of moderate size and complexity.

Problems of maintaining large amounts of data persistenly have always been the domain of competence of the database community. However, rules as a means of representing data have not been regarded as particularly important by database researchers for long time. Although relational databases have been offering the concept of a *view* since the very beginning, the relevance of intensional forms of knowledge representation is only very slowly recognized in the database community. In spite of this tendency to neglect rules, deductive databases enjoy increasing popularity today, which is mainly due to efforts of researchers trying to combine AI and database theory and technology. The foundations of the deductive database paradigm have been outlined during a serious of workshops in Toulouse between 1977 and 1982. The results of this "initialization phase" have been summarized in [GMN84], which is today regarded by many as a kind of "locus classicus" of the field. In the mid-80s three international workshops on knowledge base management and deductive databases have further promoted the interest in the subject (Islamorada 1985 [BM86], Xania 1985 [ST89], and Washington 1986 [Min88a]). Today contributions to the area of deductive databases can be found in the proceedings of all major conferences on databases as well as on logic programming. A summary of the historical evolution of deductive database research has been given in [Min88b].

The close relationship between concepts and methodology of logic programming languages and deductive databases are often misleadingly interpreted in such a way that the notion 'deductive database' is regarded as a synonym for a coupling of a PROLOG system and a database (or at least a file) system. Many research groups all over the world have been experimenting with various forms of PROLOG-DB couplings. Even some commercial PROLOG systems are already offering connections to a commercial database system. All these systems can readily claim to be called deductive databases as they offer a particular form of rules for knowledge representation and a particular kind of inference for retrieval purposes. Some of them have meanwhile reached a remarkably high standard - in particular if the file system and inference engine are closely integrated, for example like in [Boc89].

However, the concept of a deductive database is much more general and consequently has given rise to quite different interpretations. Today, the aims and ambitions of many research projects go considerably beyond the PROLOG-DB approach. The reasons for doing so are on the one hand conceptual ones, such as the desire to provide a higher degree of declarativity than offered by PROLOG. On the other hand it is performance which matters: it is well-known that the particular inference mode of PROLOG has principal weaknesses in certain cases, which become particularly awkward in presence of a large set of facts (explicitly stored or derivable). The scientific debate about the question whether database technology or logic programming language technology is more promising as a basis for rule handling is still ongoing and has not produced any definite answers yet. Consequently, one can observe a sort of fruitful competition between "PROLOG-oriented" and "database-oriented" solutions today.

There are two aspects of research in deductive databases that have received particular attention in the recent past:

- the problem of efficiently maintaining database integrity in the presence of rules
- the problem of efficiently evaluating retrieval requests over recursive rules

Most of the solutions proposed for both problems are aiming at an exploitation of database technology. A large number of articles have been (and are still) written every year addressing both topics. Many of these publications, however, are not well suited for getting acquainted with problems and solutions, as they often employ an ideosyncratic style and a heavy formalism. In this paper we try to introduce the main ideas currently investigated in advanced research in a more "popular" manner, thus addressing non-specialists in the AI as well as in the DB community who don't have time or opportunity to dive deeply into original publications and specialized terminology. In order to do so, we deliberately avoid any formal treatment of the matter. Such a style of presenting deductive databases is unusual, and may even be regarded as unscientific by some, but we hope it will be appreciated by non-specialists interested in a precise, but informal overview. In any case, we will provide selective, but extensive references to recent publications faithfully reflecting the state-of-the-art in the different topics addressed. Readers who are interested in a more elaborate and formalized treatment of the matter are thus encouraged to go beyond the necessarily limited scope of this paper.

2 Integrity Constraint Handling

The notion of *integrity* is well-established in the database field, comprising the idea that data manipulation should be done in such a way that certain semantic conditions arising from the particular area of application are respected. Such conditions are expressed by means of what are called *integrity constraints*. Static integrity constraints are a means of expressing conceptual properties of a database that are supposed to hold in every database state. Dynamic constraints define which kinds of database modifications are admissible. Constraints should, however, not be regarded as "eternal laws" designed once and for all. They may be subject to modification (as are the rules) whenever the general conception of an application changes.

Constraints are yes/no queries, i.e. Boolean queries returning either 'true' or 'false' rather than a set of facts as an answer. A static constraint can be expressed in the same query language that is used for retrieval purposes and for expressing rules. Dynamic constraints, however, may refer to past and future states of a database, thus requiring either explicit access to different states by means of additional state variables, or implicit ways of speaking about database evolution by means of temporal or dynamic operators. Such kind of facilities are usually exceeding the syntactic limits of query languages.

There has been quite intensive research on database integrity up till now, and the number of publications addressing the topic is continuously growing. This is opposed to the situation in the commercial world: many database systems don't provide integrity constraints at all, and those supporting the concept are usually not offering implementations which are satisfactory from the research point of view. If commercial systems are providing constraints at all they mostly restrict themselves to very basic and simple types of constraints, such as functional dependencies, keys, or inclusion relationships. This limitation is meanwhile critized in many areas of commercial data processing and motivates a rising interest of industry in research results, in particular in methods for handling very general kinds of constraints.

Up till now, most research contributions concerned with integrity have addressed static constraints only. There are notable exceptions, of course, like [LiS87] or [Lip88]. In connection with deductive databases investigations of integrity have exclusively focussed on static aspects. Thus in the following we will introduce aspects of static constraint handling only.

2.1   Checking Constraint Satisfaction

When handling very general constraints, the most ardent problem is the high evaluation effort associated with a complete check of all constraints after every modification of a data or knowledge base. As most integrity constraints express universal requirements, i.e., properties holding "for all" objects of a certain kind, evaluation costs often become prohibitively high. It has been observed rather early that a particular update usually affects only particular instances of particular constraints, whereas the majority of constraints remains unaffected. It has been a main concern of research since to identify such a reduced set of constraints the evaluation of which is sufficient in order to guarantee that database integrity will be preserved. By specializing a constraint set with respect to a particular update, evaluation costs can be dramatically decreased. Nevertheless, if complex constraints are arising from the application, complex specialized constraints may have to be checked despite all efforts to discard redundant computations. Such checks will remain costly and time consuming. However, there is only one very questionable way out: don't make complex constraints explicit (in order to save computations), but leave responsibility for controlled knowledge manipulation to the user. Just to distribute the necessary checks to different places in an application program does not help either, but rather constitutes a step backward into "pre-declarative" times, when knowledge was hidden in the code.

During the last decade, numerous techniques and strategies for constraint specialization have been proposed most of which can rather easily be applied in practice. Logic programming languages like PROLOG are particularly well-suited for implementing the complex symbolic manipulations required for specialization. Such a use of PROLOG for pure implementation purposes is completely independent from any particular query and constraint syntax, which may be PROLOG, but might as well be SQL or relational algebra.

In a deductive database, constraint specialization becomes a more pretentious task, as the implicit effects of updates on intensionally defined concepts have to be considered too. However, the techniques for determining implicit updates are formally identical with those for identifying and simplifying constraints affected by an update. This insight is not yet properly reflected in most of the contributions to the topic, which are still presenting both parts of the specialization process in different terms. In the remainder of this section section we will try to provide the reader with some basic ideas about constraint specialization by discussing simple, but characteristic examples. Let us start with the constraint:

"Every employee is a member of some department!".

Assuming that the current state of the database already satisfies every constraint, there is no need to re-check this constraint at all when an employee is fired. We say the constraint is not *affected* by deletions of 'employee' facts, as it still holds for those employees which remain recorded in the database after the deletion. However, the constraint might be violated in case a new employee is hired. Even in this case there is no need to re-evaluate the constraint as a whole (which might be an expensive affair, if considering the personnel database of Siemens, for example, comprising some 350,000 employees). Only the newly inserted employee (let us call him 'Jim' for the moment) has to be considered because he might not (yet) be assigned to any department. Thus only a single instance of the constraint has to be checked:

"Jim is a member of some department!"

(The evaluation effort would be reduced from 350,000 comparisons to a single look-up, in case Jim were hired by Siemens). We speak of a *relevant instance* of an affected constraint.

Affected constraints and relevant instances can be identified for entire update patterns rather than for individual updates. The fact that insertions of employees affect the example constraint is independent of the particular employee who is hired. It only depends on the constraint, the updated relation, and the "sign" of the update operation performed. Affected constraints and relevant instances can be formally defined in terms of unifiability and logical polarity.

In order to illustrate that this basic specialization step is by no means as trivial an affair as the introductory example might suggest, consider the same example constraint once more, but now assume a certain department is deleted, say the research department (not necessarily of Siemens). This deletion affects the constraint, as the former researchers have to be assigned to new departments in order to maintain integrity of the database. This time a relevant instance cannot be simply obtained by replacing the variable expression 'some department' in the constraint by the constant 'research' stemming from the update. It doesn't make any sense to check

"Every employee is a member of the research department!"

although the same instantiation step has been performed as for Jim before. The difference lies in the quantifier involved: replacing a universal expression like 'every employee' by a specific constant 'Jim' is a valid logical specialization, whereas replacing the existential expression 'some department' by the constant 'research' is an illegal generalization.

However, it is possible to exploit relevant instances of existential subexpressions of a constraint in a different manner. One may characterize those employees that are affected by the update by means of the condition "who was a member of the research department before". This condition is obtained by instantiating the existential expression by the constant in the update. It has to be necessarily satisfied by any employee for whom the constraint does not hold any more in the updated database - if any! Syntactically augmenting the constraint by adding the condition leads to a reduction of the search space during evaluation. Checking

> "Every employee who was a member of the research department before still is a member of some department now!"

involves fewer instances than checking the original constraint. However, the costs for computing the smaller set of employees may exceed the costs saved by getting rid of some instances. Whether the evaluation of the more specialized constraint pays off, or not, depends on the ratio between researchers and employees in the actual databases. In a company with few researchers it is worthwhile, in a research center evaluation of the specialized constraint will be more expensive than evaluating the original constraint anew.

The main source of variation between different methods for constraint specialization is the question whether necessary conditions like the one in our little example should be exploited, or not. Otherwise the methods proposed are essentially identical. The most significant contributions to constraint specialization in relational databases are [Nic82], [BB82], and [MH89].

Let us now turn to constraint specialization in presence of rules. As mentioned earlier, implicit updates have to be considered as well, as they might indirectly lead to constraint violations. As an example assume these two general laws:

> "Every member of some department is a properly integrated employee."
> "Every employee should be properly integrated".

Assume further that the first law is treated as a derivation rule, whereas the second one is treated as a constraint. Taken together they impose the same condition as the single constraint before: everybody has to be assigned to at least one department. However, the deletion of the research department does not directly affect the new constraint anymore. The explicit update now *induces* an indirect update, because those employees not belonging to any other department than research cease to be properly integrated. Deletions of facts for the (intensionally represented) predicate 'properly_integrated' do affect the constraint. Thus deletion of a department indirectly affects the constraint due to its implicit effect on 'properly_integrated'.

Determining such implicit consequences of updates in the presence of rules is called *update propagation*. Induced updates can be formally defined in terms of unifiability and polarity too, which doesn't come as a surprise, as the steps that have to be performed are the same regardless of whether the condition "is a member of some department" occurs explicitly in the constraint (as in the first example) or is only implicitly involved via a rule. Also the tradeoff problem mentioned (does it pay off to determe the individual members of the research department, or not) turns up again, this time as a strategical problem of update propagation. Again it depends on the actual data present whether the one or the other method of update propagation is more appropriate.

Many proposals for extending relational constraint specialization methods for deductive database purposes have been published during the last five years, such as [Dec84], [LST87], [KSS87], and [BDM88]. All these contributions (including the one by this author) suffer to various degrees from two main drawbacks:

- They do not realize that determining relevant instances and induced updates are in fact one and the same formal problem and should consequently be treated identically ([KSS87] constitutes an exception in this respect).

- They do not take into account any statistical information about the actual state of the database (like the ration between researchers and employees in the above example), but impose one immutable principle on every rule and every constraint.

Thus there is still a need for considerable improvement of the integrity checking methods both conceptually as well as strategically. Recent attempts at representing the constraint specialization process entirely by means of additional "propagation rules", thus turning the integrity checking problem into a query answering problem, may be a promising basis for progress.

2.2 Checking Constraint Satisfiability

A rather high research effort has been devoted to the problem of efficiently checking constraint satisfaction. A related problem, however, has received only very little attention up till now. Declarative languages as can be found in advanced deductive database and knowledgebase systems enable a user to express very general and complex rules and constraints. Understanding the logical consequences of such high-level forms of general knowledge is a difficult task. As a consequence it is very likely that - unintentionally - logical inconsistencies will be introduced which are of a principal nature, i.e. independent of any specific set of facts associated with rules and constraints.

Consider, for example, the following very simple and very natural constraints, taken from a hypothetical personnel database again:

1. "Every department leader is a member of the department he leads."

2. "Every member of a department works for the department leader."

3. "Nobody works for himself."

4. "Every department is led by some employee."

5. "Every employee is a member of some department."

Although each of these constraints as such seems to be extremely reasonable, it is impossible to satisfy them all together at the same time, except in a trivial database not containing any information about employees and departments at all. This may come as a surprise at first, but can be explained very easily by reasoning about the implications of the five constraints for a moment: Constraints 1 and 2 taken together imply that every department head works for himself, as he is supposed to be a member of his own department. Constraint 3 prevents anybody from working for himself. Therefore the only way the first three constraints can be satisfied is by having no department heads at all. This is only possible if there are no departments at all (due to constraint 4), which in turn implies that no employees are existing (due to constraint 5).

A simple way to retain satisfiability in this example is by adding an exception to constraint 3, for example, turning it into "Nobody works for himself, except the department leaders." It is very easy to forget or overlook such exceptions or border cases when designing even moderately complex constraints, particularly if the incompatibilities arising between contradicting requirements only become visible when several constraints are logically interacting.

Unsatisfiability is not a potential problem for integrity constraints alone. A satisfiable set of constraints may become unsatisfiable if combined with a set of rules. In our little example this can be easily understood by turning constraints 1 and 2 into derivation rules. The remaining constraints 3 to 5 are satisfiable. If the two rules are added, however, the same logical inconsistencies arise as before.

In order to detect and eliminate such logically incompatible rules and/or constraints already at design time it is necessary to apply techniques developed for automated theorem proving purposes. (A theorem prover based on refutation checks whether the negation of the theorem and the axioms from which to derive the theorem are satisfiable.) Rather than adding an arbitrary theorem prover as a separate component to a deductive database management system, however, one should make use of those inference techniques that are already available in a deductive system for query answering and constraint checking purposes.

A very elegant and convenient way of checking satisfiability while exploiting deductive database technology has been described in [BDM88] and [Man90]. The approach presented there is based on a model-generation paradigm of theorem proving introduced in [MB88]. Models of constraints and rules can be viewed as example databases in which all the constraints are satisfied. Automated model generation can thus be regarded as systematic construction of such example databases. The close resemblance between the basic paradigm of the satisfiability checker and the database context in which it is applied makes the approach particularly appealing for users.

A model can be constructed by successively inserting facts into an initially empty database. Which facts have to be inserted depends on those constraints that are still violated in the example database under construction. Each positive, atomic formula occuring in a constraint potentially requires one or more atomic insertions into the example database in order to make the respective formula true. Conjunctions and universal formulas give rise to transactions of insertions satisfying each component/instance. A model of a disjunctive or existential formula can be constructed by means of a case analysis systematically considering the different alternative components/instances of the formula. Completely negative constraints (like "Nobody works for himself") rule out certain combinations of facts. In case all attempts of constructing a model fail, unsatisfiability has been shown. Again PROLOG turns out to be an ideal implementation tool for a model generation procedure, as the language provides the necessary control structures for case analysis, failure, and backtracking.

The undecidability of the satisfiability problem is a principle obstacle that cannot be overcome, except by unacceptably severe syntactical restrictions. However, termination of the model generation process can be guaranteed for the vast majority of cases. Only if the designer has formulated constraints that cannot be satisfied in any finite database model generation of course does not stop, as it faithfully reflects what the designer has required. Again it is very easy to unintentionally come up with such undesirable constellations of constraints (and rules) by just forgetting certain exceptions. For a more detailed discussion of the model-generation approach refer to the literature cited in this section.

2.3 Intensional Answers and Semantic Query Optimization

Another problem related to the notion of integrity constraints is also just beginning to be mastered by research. Whereas up to now integrity constraints have only been used for controlling database modifications, it is possible and desirable to exploit the general knowledge expressed as constraints for query answering purposes as well. As opposed to rules, most constraints cannot be used for implicit representation of facts, as they often express negative or indefinite knowledge. However it is possible to generate non-factual, *intensional answers* by means of constraints. Moreover, constraints can be used for internally controlling and optimizing the evaluation of queries. The term *semantic optimization* has been introduced in this context.

In order to clearly understand the key idea behind the notion of an intensional answer consider the query

"Which employees earn less than $50,000?".

An ordinary database system would answer such a retrieval request by listing the names of all those employees that satisfy the condition. Imagine there is a constraint

"No employee under 30 earns more than $40,000!"

An intensional answer to the query would consist of the general formula

"All employees under 30 earn less than $50,000".

If the user is interested in a more detailed, extensional form of the answer, he may subsequently ask for those employees who are under 30. The rationale behind the intensional answer concept is, however, that in many cases general qualifications of the set of answers are much more meaningful to the user than an exhaustive listing of individual answers. One of the main problems with intensional answers is the question of completeness. In our example, there might be employees older than 30 as well who earn less than $50,000. How is it possible to conveniently and efficiently characterize the remaining answers not covered by the intensional description? There are only a few serious contributions to the problem up till now. The solutions proposed are not going very far yet. In order to get an impression of the state-of-the-art in intensional answer generation refer to [Imi87], [CD88], [SM88], [Mot89], and [PR89].

The problem of semantic query optimization is closely related to intensional answering, as integrity constraints are exploited during the query answering process in both cases. Again we illustrate the basic principle of semantic optimization by means of a simple, characteristic example. Consider now a very similar example query:

"Which employees earn more than $50,000?"

Assume this time that the concept 'employee' is intensionally defined by means of three rules, reflecting that employees can be either managers, or engineers, or secretaries. Evaluation of the example query will thus consist of answering three independent subqueries, one for each rule defining 'employee'. Assume furthermore that there are two constraints:

"Secretaries cannot earn more than $30,000!"
"Only engineers with a doctoral degree may earn more than $40,000!"

The first constraint can be used to discard the first subquery immediately: there are no secretaries satisfying the query qualification. Thus any search can be avoided in this case. The second constraint can be used for reducing the search space by specializing the second subquery into

"Which engineers with a doctoral degree earn more than $50,000?"

As far as proposals of general techniques for semantic optimization is concerned, the situation to date is similar to the situation of intensional answer generation: there are a number of initial studies in various formal systems, but no convincing, generally acknowledged solution has emerged yet. As introductory literature we recommend [SO87] and [CGM88].

3 Recursive Rule Handling

Many "traditional" database researchers still regard recursion as not much more than a curiosity. If considering the matter a little bit more seriously and without prejudices, however, one will soon understand that recursive rules are by no means just "exotic" border cases which don't appear in the "real" world. Whenever knowledge is represented declaratively in the context of hierarchic structures, recursion is inevitable. Identifying components of a complex design object, finding connections in a timetable or on a road map, computing the total cost of a composite part in an engineering database (the famous 'bill of material' problem): recursive rules are the simplest and most natural way of declaratively representing relationships in the graph-like structures underlying all these applications. Even if it should turn out that most recursive problems are in fact variations of a transitive closure computation: not to be able to represent recursively defined concepts declaratively constitutes a serious drawback of any knowledge representation system, be it an AI system or a database system. Being forced to encode recursion procedurally, e.g., by calling the Warshall algorithm whenever a transitive closure is occuring, can only be regarded as an "emergency exit", but not as a serious solution.

In relational database systems recursive view definitions are nearly always excluded, as the standard technique of view handling (complete expansion of all view definitions before evaluation) does not work in presence of recursion. Even if recursive rules are syntactically admitted, the majority of today's rule-based systems is not able to always guarantee termination and completeness of the query evaluation process. Systems implemented in PROLOG, e.g., do not terminate when evaluating recursive rules over cyclic structures. Simple iterative solutions often found in expert systems suffer from severe inefficiency as they are not able to make use of the constants in a query for reducing the search space.

The problem of developing an efficient evaluation mechanism able to handle any kind of rules correctly and completely has been one of the main areas of activity in deductive database research from the very beginning. The majority of approaches has investigated the problem of recursion in connection with a set-oriented evaluation paradigm, as usual in a database context. Although logic programming languages and expert systems are dealing with recursion in a setting where individual answers rather than sets of answers are computed, the principal problems caused by recursion remain basically the same. The discussion about the most appropriate way of overcoming these problems often tends to be obscured by the more general discussion about the most promising technique for handling rules in general. Particularly in the database community there is an ongoing debate about the question whether deductive query evaluation methods should be based on bottom-up or on top-down reasoning. One of the most prominent researchers in the field has recently published a paper which already in its title announces that "bottom-up beats top-down" [Ull89b]. This claim - which pretends to provide an answer to the debate mentioned - is based on a very special and restricted top-down method, however. Therefore it is not particularly well suited for establishing a general statement about the one or the other paradigm.

Moreover, in a very recent analysis of the "competing" approaches ([Bry89]) it has been convincingly demonstrated that those optimizations that have been proposed for bottom-up evaluation in fact implement a special form of top-down reasoning which guarantees termination and answer-completeness for recursive rules. The modifications of the basic top-down paradigm achieved this way turn out to be the same as those independently proposed by researchers working solely in a top-down framework. [Bry89] shows that the two most widely known representatives of each "camp" - the magic set approach ([BMSU86],[RLK86]),[BR87]) and the QSQ approach ([Vie86],[Vie89],[LV89]) - can be very elegantly explained as two different implementations of a more general principle. Those solutions for the recursion problem that have been proposed in a logic programming framework - such as [TS86] or [Die87] - find a natural explanation in Bry's framework, too.

In spite of this unified view, which has significantly contributed to a better understanding of rule handling in general and recursion in particular, the matter remains intricate and difficult to present. In the following we will therefore try to give the reader at least an idea of what has been achieved to date with respect to recursive rule handling. Such a presentation will necessarily remain incomplete and sketchy, but hopefully makes it possible to understand a little bit why this topic has attracted so many researchers up till now, and why its results still look like "magic" to many. For those interested in a more detailed presentation of the topic we recommend two recent textbooks a considerable part of which is devoted to an in-depth presentation of rule handling strategies in deductive database: [Ull89a] and [CGT90].

3.1 Handling recursive rules by bottom-up iteration

In the database community, the predominant framework for investigating recursion has been the bottom-up paradigm up till now. This tendency can be explained by the very straightforward way how this approach can be implemented by means of traditional relational database technology. But also for production rule systems and for explaining the foundations of logic programming ([vEK76]) this paradigm has been chosen. In this section we will discuss (again by means of an example) in how far recursion represents a problem for bottom-up methods.

First, however, let us shortly recall the basic principle of this approach. As a running example we will use a simple transitive closure definition, representing paths in a graph by means of a non-recursive and a recursive rule:

"Every edge is a path."
"A path can be obtained by connecting an edge with a path."

In order to facilitate the discussion of some more detailed points, we will depart from our natural language style of expressing rules in the following. We will use a moderately formal representation (closely resembling logic programming conventions) instead, using predicates and variables. In this style a transitive closure definition looks as follows:

path(X,Y) **if** edge(X,Y)
path(X,Y) **if** edge(X,Z) **and** path(Z,Y)

We assume that 'edge(X,Y)' reads "there is an edge from node X to node Y". We will use these two rules in connection with a small example graph represented in terms of the following facts stored extensionally in the 'edge' relation:

edge(a,b)   edge(b,c)
edge(d,e)   edge(b,d)

The notion 'bottom-up' is motivated by a representation of query evaluation over rules by means of a derivation tree. The query represents the root of the tree, i.e., its top node. Leaves of the tree correspond to extensionally stored (or "base") facts. Intermediate nodes represent derivable facts which are needed as intermediate results for obtaining answers. The edges in a derivation tree correspond to individual derivation steps. The derivation tree representation very naturally leads to a paradigm where rules are exclusively used as generators for intensionally defined data. By evaluating the body of a rule over the base facts and over those derived facts that have already been computed before, bindings for the variables in the head of the rule are obtained, resulting in a new derived fact.

The basic strategy for bottom-up processing is a simple iteration scheme: Initially, all rules are applied to all base facts. The resulting derived facts are temporarily stored in order to be available as input for further derivations. Then the rules are applied anew, taking into consideration the "first generation" of derived facts as well, which may lead to a "second generation", and so on. Iterated application of forward reasoning terminates as soon as in one "round" no new derived facts can be obtained anymore. For our example graph iteration looks as follows:

1. Initially only the non-recursive rule is applicable, thus all paths of length 1 are computed and stored: path(a,b), path(b,c), path(b,d), path(d,e).

2. The non-recursive rule does not contribute any new facts anymore, but the recursive rule becomes applicable as facts for 'path' are available now. All paths of length 2 are computed: path(a,c), path(a,d), path(b,e)

3. The only new fact generated corresponds to the only path of length 3: path(a,e)

4. All paths have been computed, iteration stops.

A mathematical formalization of such a process in terms of fixpoint theory is well-established in computer science. Therefore this approach is often called bottom-up fixpoint procedure. Once a fixpoint has been computed, the answers for the particular query to be answered can be extracted from the temporarily stored derived facts.

Two technical details of bottom-up processing deserve special attention, in particular in connection with recursion. In order to know when to stop the iteration process one has to determine for each fact generated during an iteration round whether it is actually new, or has already been derived before and thus has already been temporarily stored. This *duplicate check* is a basic prerequisite for ensuring termination. A second particularity is the temporary *materialization* of the derived facts generated during each iteration round. This step is necessary because the output of previous rounds has to be used as input for future rounds, at least in those cases where the body of a rule (not necessarily a recursive one) contains a reference to some intensionally defined concept.

Both, materialization of derived facts as well as duplicate check, can be implemented rather easily using conventional database technology. Temporary relations can be used for storing intermediate results. Relational operators like join or union filter out duplicates anyway, as they are supposed to return sets, which are duplicate-free. Thus bottom-up evaluation reduces to a process of iterated query answering over temporary and base relations benefiting from optimization techniques developed for non-deductive applications.

3.2   Top-down optimizations

In spite of its simplicity and its closeness to established database technology, the bottom-up paradigm in its pure form suffers from a serious drawback. It does not take into account the particular query to be answered. For each query, all intensionally defined facts are computed in order to make sure that all answers are found. This is heavily redundant as in most cases only a few rules are needed for deriving a particular (set of) answer(s). In addition, if a query is partly or fully instantiated, only a subset of all derivable facts is involved in possible derivations of answers.

In our example, the first problem - firing irrelevant rules - does not arise, as we only have a single intensional concept. But it is easy to imagine what would happen if our two rules were part of a huge knowledge base! It is a matter of course that bottom-up query answering necessarily requires to determine those rules which are potentially needed for answering a particular query in advance. This can be achieved for each type of query by means of a simple analysis performing a symbolic top-down expansion of the respective predicate.

The second problem - generation of irrelevant facts - becomes visible already in our tiny toy example. Suppose we wanted to know all nodes reachable from 'b' in our example graph, i.e., we pose the query 'path(b,Y)'. In order to generate the three answers 'path(b,c)', 'path(b,d)', and 'path(b,e)' we only need one additional intermediate result - 'path(d,e)'. The simple iteration scheme described above would nevertheless compute all paths in the graph, including the four paths starting at 'a', which are completely irrelevant for answering the query. In our example it doesn't really matter whether four or eight facts are generated. In a big realistic example, however, the inability of simple bottom-up reasoning to avoid redundant computations would easily result in an enormous overhead, ruling out this approach for any reasonably complex application.

Being able to reduce the search space during query evaluation is crucial. One of the most important optimization steps in relational databases aims at achieving such an effect by trying to push selections (i.e., conditions restricting the values of certain variables) as deeply as possible into each query in order to avoid generation of irrelevant intermediate results as early as possible. Applied to rule-based query evaluation this kind of optimization would correspond to another form of top-down analysis ahead of actual evaluation. This time top-down expansion of the query is done in order to successively instantiate as many variables as possible by means of constants from the query. The intention is to use rules top-down as a means of problem reduction before applying them bottom-up for answer generation. In our example, it is possible to push the constant 'b' appearing in the query 'path(b,Y)' into both rules defining 'path' by means of a one-level expansion, thus obtaining two more specialized rules:

path(b,Y) **if** edge(b,Y)
path(b,Y) **if** edge(b,Z) **and** path(Z,Y)

Applying the first of these rules yields the two answers 'path(b,c)' and 'path(b,d)' immediately. However, applying the second rule does not produce any further answers: 'path(b,e)' is lost! This is due to the fact that for computing the third answer we need 'path(d,e)' as an intermediate result to be conjoined with 'edge(b,d)'. As we have specialized the rules for answering 'path(b,Y)', we have lost the ability to answer the recursive subquery 'path(d,Y)'. Thus, pushing selections into rules statically, i.e., ahead of evaluation, may lead to an over-specialization of rules resulting in incompleteness. However, problems of this kind only arise for recursive rules. In case all rules are non-recursive, a static top-down analysis and specialization of rules works perfectly well in combination with bottom-up iteration. This strategy has also been used for implementing views in relational databases. The fact that it cannot be applied in the recursive case is the main reason why recursive views have not been admitted.

What is needed in order to reduce the search space in presence of recursion is a form of dynamic propagation of selections applying specialization steps to different "incarnations" of the recursive rule, very similar to the way how evaluation in PROLOG would proceed. In PROLOG the recursive rule would be applied as follows: The top query 'path(b,Y)' is expanded resulting in a first incarnation of the recursive rule which is specialized by instantiation of 'X':

path(b,Y) **if** edge(b,Z) **and** path(Z,Y)

Before expansion continues, however, an answer is generated for the first of the two resulting subqueries 'edge(b,Z)' and the rule is further instantiated by means of the resulting binding for 'Z' (in order to keep things reasonably short we immediately turn to the second answer 'edge(b,d)'; 'edge(b,c)' is treated similarly):

path(b,Y) **if** edge(b,d) **and** path(d,Y)

This way the second subquery has become more specific as well. Now top-down expansion of 'path(d,Y)' takes place creating its own specialized incarnation of the rules defining 'path':

path(d,Y) **if** edge(d,Y)
path(d,Y) **if** edge(d,Z) **and** path(Z,Y)

Applying the non-recursive rule for answer generation yields 'path(d,e)' (the intermediate result missing in the static case!) which is immediately passed bottom-up to the top query resulting in the missing answer 'path(b,e)'. Application of the non-recursive rule starts again by an evaluation of the 'edge' subquery which leads to a further specialization of this incarnation too:

path(d,Y) **if** edge(d,e) **and** path(e,Y)

The newly generated recursive subquery, 'path(e,Y)', is now treated accordingly. The resulting incarnation of the non-recursive rule does not contribute any new answer because 'e' is a leaf in our graph. The incarnation of the recursive rule does not lead to any further expansion for the same reason: no answer to 'edge(e,Z)' can be found anymore. Thus a PROLOG-like evaluation of our example query produces all answers and does not generate any unnecessary intermediate result. The "key to success" is PROLOGs dynamic expansion of (sub)queries creating as many different incarnations of recursive rules as required. In addition, PROLOG interleaves top-down expansion steps and bottom-up generation steps very closely thus being able to instantiate subqueries as much as possible before expanding them.

However, also the PROLOG strategy is not perfect. It fails whenever cyclic graphs are involved. Assume we didn't have 'edge(d,e)', but 'edge(d,b)' in our graph, leading to a cycle $b \Rightarrow c \Rightarrow d \Rightarrow b$. In this case our last incarnation of the recursive rule looks different:

path(d,Y) **if** edge(d,e) **and** path(b,Y)

The new recursive subquery 'path(b,Y)' is identical to the top query from which we have started. PROLOG implementations do not detect such repetitions. Therefore the whole process starts anew and continues "ad infinitum". Note that this kind of looping is a particularity of PROLOG (motivated by implementation problems), not of top-down reasoning in general. A similar pathological behaviour would arise in the bottom-up case, if the duplicate check were omitted.

Back to optimizing bottom-up iteration! The ambition of researchers working on this topic was to come up with an optimized version of bottom-up iteration which was able to apply a dynamic propagation of restricting constants like the one achieved by PROLOG without giving up the convenient iteration paradigm and without "inheriting" the shortcomings of PROLOG. Some five years ago, two proposals were published practically at the same time, which achieved this goal by means of the same techniques (only differing in terminology and in minor details): the Alexander method [RLK86] and the Magic Set method [BMSU86]. Whereas the magic set approach became very quickly very popular, the (identical) Alexander method received far less attention.

3.3 Magic rewriting

The key to understanding the magic set approach is the observation that a top-down expansion of queries can be formalized in terms of fixpoint iteration as well. Instead of answers and intermediate results, subqueries are generated level by level by iteratively applying rules until all different possible subqueries have been obtained[1]. Such a process can be described (on a meta-level!) by means of artificial rules as well, which are then treated by means of bottom-up iteration.

Because this very unconventional way of viewing top-down expansion is the most "magical" aspect of the magic set approach, we will illustrate it with the help of our little example. How are subqueries generated in the transitive closure case? As queries refering to the base predicate 'edge' do not present any problems, we concentrate on queries to the 'path' relation. Subqueries for 'path' are only obtainable through the recursive rule:

path(X,Y) **if** edge(X,Z) **and** path(Z,Y)

In order to reach a subquery 'path(Z,Y)' during top-down expansion, two conditions have to be satisfied:

1. A query 'path(X,Y)' has been posed before, responsible for the creation of the respective incarnation of the rule.

2. The subquery 'edge(X,Z)' has successfully returned bindings for 'X' and 'Z'.

Assume that subqueries already generated have been temporarily stored in an auxiliary (meta) relation 'query'. Now we can represent the two conditions for subquery generation in terms of an auxiliary (meta) rule defining the extension of 'query':

query(path(Z,Y)) **if** query(path(X,Y)) **and** edge(X,Z)

This rule is recursive. In order to be applicable at all we need a non-recursive rule defining 'query', or at least one initial 'query' fact. Such a fact is obtained by initially putting the top query into the 'query' relation. In our example we start from 'query(path(b,Y))'. Now let us apply bottom-up iteration to the auxiliary rule and the auxiliary fact:

---

[1]Note that we are adopting a typical restriction in deductive databases, namely to exclude function symbols. Thus only finitely many different subqueries can arise.

1. Initially two new 'query' facts are generated: 'query(path(c,Y))' and 'query(path(d,Y))'

2. Now the facts generated in the first round are exploited as input: 'query(path(c,Y))' does not lead to any new result, because there is no fact matching 'edge(c,Z)', but 'query(path(d,Y))' contributes to the generation of 'query(path(e,Y))'.

3. No new 'query' fact can be found anymore, because 'query(path(e,Y))' cannot be combined with any 'edge(e,Z)' fact. Thus iteration stops after the third round.

The fixpoint computed this way consists of the following four auxiliary facts:

query(path(b,Y))	query(path(c,Y))
query(path(d,Y))	query(path(e,Y))

These are exactly those four subqueries generated by a PROLOG-like evaluation as described above. Top-down expansion of queries has been implemented by means of a bottom-up iteration over the auxiliary rule!

However, our goal was not only to describe query expansion, but to achieve a restriction of bottom-up answer generation by propagating selections. This goal can be reached by modifying each of our original rules:

$$path(X,Y) \text{ if } query(path(X,Y)) \text{ and } edge(X,Y)$$
$$path(X,Y) \text{ if } query(path(X,Y)) \text{ and } edge(X,Z) \text{ and } path(Z,Y)$$

Each rule body has been prefixed with a reference to the 'query' relation, which can be understood as a kind of "guard": rules can only be applied for answer generation now, if previously a corresponding (sub)query has been generated. In addition, the added 'query' references have the effect that bindings for those variables occuring in the head of the rule become available during evaluation of the body, thus providing the same specialization as obtained during top-down expansion.

Let us illustrate this "miraculous" effect with our little example again. How does bottom-up iteration of the "magically" rewritten 'path' rules proceed, assuming we have already generated the respective 'query' meta facts before:

1. Initially only the non-recursive rule can be applied for bindings 'X=b', 'X=c', 'X=d', and 'X=e', resulting in the generation of 'path(b,c)', 'path(b,d), and 'path(d,e)'.

2. Now the recursive rule can be activated, resulting in the generation of a single new fact: 'path(b,e)'

3. A third round doesn't produce any new facts: iteration terminates.

The fixpoint of this iteration has produced exactly what is required: the three answers to the initial query and the single intermediate result 'path(d,e)'. Thus we have reached our goal to come up with an optimized bottom-up iteration that does not produce any irrelevant intermediate results. Let us summarize what has happened:

1. In addition to generating and storing answers and intermediate results, the bottom-up principle has been applied to subqueries as well.

2. Subquery generation is achieved by means of additional auxiliary rules expressing exactly how subqueries are generated during top-down expansion.

3. The original set of rules are modified in such a way, that they take the subqueries generated into account, thus preventing irrelevant activations of rules. In addition, the constants contained in the subqueries are exploited for specializing the rules used for answer generation, thus preventing generation of irrelevant facts.

4. The top query to be answered is encoded by means of an initial 'query' fact, which is "fed" into the evaluation process at the very beginning. By this encoding the top query in the original evaluation process becomes a bottom fact during evaluation of the modified rules.

The most important point concerning the evaluation of "magically rewritten" rules is that although a bottom-up iteration has been retained as the underlying mechanism, a completely different sequence of individual processing steps has been achieved (if compared with the original, unmodified set of rules). In fact, the individual steps of this optimized fixpoint computation correspond one-to-one to the steps of a set-oriented top-down evaluation performed in a breadth-first manner. This is the reason why the simultaneous fixpoint computation of subqueries and answers has been called 'top-down fixpoint procedure' in [Bry89]. Bottom-up iteration over the rewritten rule system "implements" top-down evaluation with respect to the original rules. In order to convince himself of this fact, the reader is encouraged to go through the example once more himself, this time iterating over both kinds of rules, those for 'query' and those for 'path' simultaneously, and to compare each step with the corresponding top-down step.

How does the 'top-down fixpoint procedure' avoid the shortcoming of PROLOG in case of cyclic data? The answer is provided by the duplicate check performed during bottom-up iteration: Whenever a fact has been generated during an iteration round, it is checked whether this fact is already contained in the respective temporary relation. As subqueries are represented as facts as well, loop checking is achieved as a kind of "by-product".

When trying to understand the rationale behind the magic rewriting approach, one might ask the question, whether it is really a good idea to implement query evaluation in such a "funny" way. However, attempts to address the problem purely in terms of top-down reasoning - like in Vieille's QSQ-approach - have led to very similar results. In order to make top-down reasoning terminating and answer-complete in a set-oriented framework, three modifications of the basic, PROLOG-like evaluation scheme had to be provided:

1. In order to implement the loop check, an intermediate storage for subqueries was needed. As the amount of queries to be managed can be quite huge in a database context, a representation in a main-memory stack (like in PROLOG implementations) had to be given up, and relational technology had to be used.

2. In order to retain answer-completeness, answers and intermediate results for recursive subqueries had to be stored in temporary relations as well. Exploitation of these stored results is achieved by "feeding them back" into the evaluation process as base facts for those recursive subqueries that have not been further expanded because of the loop check.

3. Finally, the depth-first strategy of PROLOG had to be given up, because there are cases where even loop checking and "feed back" of recursive answers does not guarantee completeness (a standard example for this problem has been given in [Nej87]). The modified strategy is essentially a breadth-first one.

Even the resulting implementations developed for enhanced top-down processing (such as the one described in [LV89]) have finally turned out to closely resemble a bottom-up iteration over magically rewritten rules.

At the end of this short "excursion" into the "magical" world of recursion, we will explain a technical point that might facilitate understanding for those readers who like to understand in more detail, what is behind the magic set approach. In the papers describing the method, the idea that the additional auxiliary relations in fact contain representations of subqueries generated top-down is not always made explicit. Instead, a whole lot of different 'magic' relations are introduced, one for each recursive predicate and each instantiation pattern (called "adornment") in the original set of rules. These many different auxiliary relations can be derived from the single 'query' relation we have chosen in a very simple and systematic manner. In our example, the respective rules in the terminology of the magic set method would be:

$$\text{magic.path.1}(Z) \textbf{ if } \text{magic.path.1}(X) \textbf{ and } \text{edge}(X,Z)$$

This different form of encoding has been chosen in order to avoid having nested structures containing variables as arguments of a fact. The method has been designed in order to exploit relational database technology, which does not handle non-atomic values. Instead of having a nesting 'query(path(..,..))' a separate auxiliary predicate 'magic.path' is used. In order to avoid variables - such as 'Y' in 'query(path(b,Y))' - only the instantiated argument is retained and its position in the original argument list is indicated by attaching a position number to the auxiliary predicate, in this case 'magic.path.1'. Once this confusing encoding (which only serves implementation purposes) is replaced by the much simpler 'query' style, most of the "magic" flavour is gone!

4 Conclusion

In this paper we have tried to outline some of the main problems and some of the key results in two areas of research in deductive databases: integrity constraint handling and recursive query evaluation. Each of these topics has received strong attention in the research community up till now. In spite of a large amount of contributions published, the relevance of the problems addressed and solutions proposed has not yet been acknowledged accordingly. This might be due to the fact that most contributions are hard to read and understand as a consequence of their heavily formal style. By giving a presentation which is consciously a very informal one, we hope to reach a wider audience. In particular we aim at non-specialists in the AI as well as in the DB community, who might become interested in deductive database research once they have had the opportunity to enter the field without too high initial barriers.

Acknowledgements

The author would like to thank his colleagues François Bry, Jesper Larsson Traff and Sury Sripada for helpful comments on earlier drafts of this paper, and Dimitris Karagiannis for his insisting patience.

References

[BB82]   P. Bernstein and B. Blaustein, "Fast methods for testing quantified relatio-
           nal calculus assertions", in: Proc. 8th ACM-SIGMOD Conf. on Management
           of Data, Orlando, 1982

[BDM88]   F. Bry, H. Decker, and R. Manthey, "A uniform approach to constraint
           satisfaction and constraint satisfiability in deductive databases", in: Proc.
           1st Intern. Conf. on Extending Database Technology (EDBT), Venice, 1988

[BM86]   M. Brodie and J. Mylopoulos [eds.], "On Knowledge Base Management Sys-
           tems", Springer, 1986

[BMSU86]   F. Bancilhon, D. Maier, Y. Sagiv, and J. Ullman, "Magic sets and other
           strange ways to implement logic programs", in: Proc. 5th ACM Symp. on
           Principles of Database Systems (PODS), 1986

[Boc89]   J. Bocca et.al., "KB-Prolog: A Prolog for very large knowledge bases",
           in: Proc. 7th British National Conf. on Databases (BNCOD), 1989

[BR87]   C. Beeri and R. Ramakrishnan, "On the power of magic", in: Proc. 6th
           ACM Symp. on Principles of Database Systems (PODS), 1987

[Bry89]   F. Bry, "Query evaluation in recursive databases: bottom-up and top-down
           reconciled", in: Proc. 1st Intern. Conf. on Deductive and Object-Oriented
           Databases (DOOD), Kyoto, Dec. 1989 (also in: Data and Knowledge Engi-
           neering, Vol. 5, No. 4, Oct. 1990, pp. 289-312)

[CD88]   F. Cuppens and R. Demolombe, "Cooperative answering: a methodology to
           provide intelligent access to databases", in: Proc. 2nd Intern. Conf. on Ex-
           pert Databases (EDS), Tysons Corner, 1988

[CGM88]   U. Chakravarthy, J. Grant, and J. Minker, "Foundations of semantic query
           optimization for deductive databases", in: [Min88a]

[CGT90]   S. Ceri, G, Gottlob, and L. Tanca, "Logic programming and databases",
           Springer, 1990

[Dec84]   H. Decker, "Integrity enforcement on deductive databases", in: Proc. 1st
           Intern. Conf. on Expert Database Systems (EDS), Charleston, 1986

[Die87]   S.W. Dietrich, "Extension tables: memo relations in logic programming",
           in: Proc. Symp. on Logic Programming (SLP), San Francisco, 1987

[GMN84]   H. Gallaire, J. Minker, and J.-M. Nicolas, "Logic and databases: a deductive
           approach", ACM Computing Surveys, Vol. 16, No. 2, 1984

[Imi87]   T. Imielinski, "Intelligent query answering in rule based systems", Journal
           of Logic Programming, Vol. 4, No. 2, 1987

[KSS87]   R. Kowalski, F. Sadri, and P. Soper, "Integrity checking in deductive data-
           bases", in: Proc. 13th Intern. Conf. on Very Large Databases (VLDB),
           Brighton, 1987

[Lip88]   U. Lipeck, "Transformation of dynamic integrity constraints into transaction
           specifications", in: Proc. 2nd Intern. Conf. on Database Theory (ICDT),
           Bruges, 1988 (also in: Theoretical Computer Science, Vol. 76, No. 1, 1990)

[LiS87]   U. Lipeck and G. Saake, "Monitoring dynamic integrity constraints based on
           temporal logic", Information Systems, Vol. 12, 1987

[LST87]   J. Lloyd, E. Sonenberg, and R. Topor, "Integrity constraint checking in stra-
           tified databases", Journal of Logic Programming, Vol. 4, No. 4, 1987

[LV89]   A. Lefebvre and L. Vieille, "On deductive query evaluation in the Dedgin*
           system", in: Proc. 1st Intern. Conf. on Deductive and Object-Oriented Data-
           bases (DOOD), Kyoto, Dec. 1989

[Man90]   R. Manthey, "Satisfiability of integrity constraints: Reflections on a neglected
           problem", in: Proc. 2nd Intern. Workshop on Foundations of Models and
           Languages for Data and Objects, Aigen, 1990 (available from TU Clausthal)

[MB88]   R. Manthey and F. Bry, "SATCHMO: A theorem prover implemented in
           Prolog", in: Proc. 9th Intern. Conf. on Automated Deduction (CADE),
           Chicago, 1988

[MH89]   W. McCune and L. Henschen, "Maintaining state constraints in relational
           databases: A proof theoretic basis", Journal of the ACM, Vol. 36, No. 1, 1989

[Min88a]   J. Minker [ed.], "Foundations of deductive databases and logic programming",
           Morgan Kaufmann, 1988

[Min88b]   J. Minker, "Perspectives in deductive databases", Journal of Logic Program-
           ming, Vol. 5, No. 1, 1988

[Mot89]   A. Motro, "Using integrity constraints to provide intensional answers to
           relational queries", in: Proc. 15th Intern. Conf. on Very Large Databases
           (VLDB), Amsterdam, 1989

[Nej87]   W. Nejdl, "Recursive strategies for answering recursive queries - the RQA/
           FQI strategy", in: Proc. 13th Intern. Conf. on Very Large Databases (VLDB),
           Brighton, 1987

[Nic82]   J.-M. Nicolas, "Logic for improving integrity checking in relational databases",
           Acta Informatica, Vol. 18, No. 3, 1982

[PR89]   A. Pirotte and D. Roelants, "Constraints for improving the generation of in-
           tensional answers in a deductive database", in: Proc. 5th IEEE Intern. Conf.
           on Data Engineering, Los Angeles, 1989

[RLK86]   J. Rohmer, R. Lescoeur, and J.-M. Kerisit, "The Alexander method: A tech-
           nique for the processing of recursive axioms in deductive databases", New
           Generation Computing, Vol. 4, No. 3, 1986

[SM88]   C.-D. Shum, R. Muntz, "Implicit representation of extensional answers",
           in: Proc. 2nd Intern. Conf. on Expert Database Systems (EDS), Tysons
           Corner, 1988

[SO87]   S. Shenoy and Z. Oszoyoglu, "A system for semantic query optimization", in:
           Proc. 13th ACM-SIGMOD Conf. on Management of Data, San Francisco, 1987

[ST89]   J. Schmidt and C. Thanos [eds.], "Foundations of Knowledge Base Management",
           Springer, 1989

[TS86]   H. Tamaki and T. Sato, "OLD resolution with tabulation", in: Proc. 3rd
           Intern. Conf. on Logic Programming (ICLP), London, 1986

[Ull89a]   J. Ullman, "Principles of database and knowledge base systems", Vol. 2,
           Computer Science Press, 1989

[Ull89b]   J. Ullman, "Bottom-up beats top-down for Datalog", in: Proc. 8th ACM
           Symp. on Principles of Database Systems (PODS), Philadelphia, 1989

[vEK76]   M. van Emden and R. Kowalski, "The semantics of predicate logic as a pro-
           gramming language", Journal of the ACM, Vol. 23, No. 4, 1976

[Vie86]   L. Vieille, "Recursive axioms in deductive databases: The Query-Subquery
           approach", in: Proc. 1st Intern. Conf. on Expert Database Systems (EDS),
           Charleston, 1986

[Vie89]   L. Vieille, "Recursive query processing: The power of logic",
           Theoretical Computer Science, Vol. 69, No.1, 1989

An Approach to DBS-based Knowledge Management

Nelson Mattos

University of Kaiserslautern
Department of Computer Science
P.O. Box 3049, D - 6750 Kaiserslautern, West Germany
e-mail:mattos@informatik.uni-kl.de

Abstract

This paper discusses the requirements of knowledge management from the point of view of Database Systems. Primarily, it focuses on the practical investigations that led to a DBS-based architecture for an effective and efficient management of large knowledge bases. This architecture, called KRISYS, is described, thereby concentrating on the mechanisms used to support the nearby application locality concept as well as to map knowledge structures onto secondary storage.

1. Introduction

Motivation

The acceleration of Knowledge-based Systems (KS) development in the last few years has revealed the enormous possibilities of the field for potential applications, rapidly stimulating the introduction of KS in the marketplace. During this introduction, some applications have indicated the feasibility of the field, whereas many others showed that their requirements lie beyond the current technology. Current limitations, which arise above all from the growing demands on managing knowledge bases (KB) appropriately, require the following [Ma89]:

- efficient management of large KB

  Large KB exceed virtual memory sizes, so that they have to be maintained on secondary storage devices, resulting in a need for adequate management and efficient access techniques.

- adequate multi-user operation

  In real world applications, KS together with the underlying KB are shared by several users. As such, multi-user facilities have to be offered in order to avoid redundancy in stored knowledge and consistency problems caused when parallel changes of the KB are allowed.

- suitability for server/workstation environments

  Nowadays, powerful workstations offer the interactive computing features required by KS. However, KS running on such workstations and using a centralized KB maintained in a remote server exhibit very low performance due to the communication and transfer overhead between the two machines involved in this environment.

- high reliability and availability

  In order to be applicable in a production environment, KS must guarantee high availability and reliability (i.e., that their functions are performed correctly) even if the system crashes or a loss of storage media interrupts their operation in an uncontrolled fashion.

- new inference mechanisms

  Existing inference mechanisms have been designed to work with KB kept in virtual memory. When used for large KB resident on secondary storage, they are computationally intolerable.

- distribution of the knowledge base

  KS for large applications involving several geographically dispersed locations should provide facilities for distributing the KB across multiple machines in order to guarantee efficiency of processing and increased knowledge availability.

- knowledge independence

  Most present-day KS are knowledge representation dependent. This means that the way in which the KB is organized and the way in which it is accessed are both dictated by the requirements of the KS and, moreover, that the information about the organization and access technique is built into the KS logic (i.e., embedded in their program's code). In such KS, it is impossible to change the storage structures or access strategies without affecting them, more than likely drastically.

These new requirements lead today to similar considerations as those made 25 years ago as the demands on data management tasks of commercial applications began to become more complex. At that time, people realized that these demands were very similar in all applications so that the most reasonable solution was to efficiently implement data management tasks only once in a separate system, removing them from the application programs and delegating them to this independent new system, thereby resulting in the emergence of what we today call Data Base Management System (DBMS) [Da83,Ul82,LS87].

Like the solution found for commercial applications, the answer to the existing KS limitations is also the delegation of knowledge management tasks to an independent system, where they are achieved in an efficient and adequate way. Analogous to DBMS, such a system is called Knowledge Base Management System (KBMS) [BM86,ST89].

Knowledge management requirements

Systems for knowledge management should integrate three different classes of functions which support the means for

- modeling the knowledge involved with a particular application, that is, constructing a KB,

- knowledge manipulation, i.e., exploiting KB contents to solve problems, and

- knowledge or KB maintenance, e.g., storage mangement, efficiency, etc.

Clearly, such systems can only support KS operations that are application and problem class independent. That is, the support of all functions provided by the KB and inference engine components, i.e., those associated with KR schemes. Problem solving strategies [St82,Pu86b,Pu88] as well as KS interfaces are closely related to the characteristics of the real world applications and their problem classes and are, for this reason, to be kept within the KS.

In this paper, we first investigate the requirements of knowledge management from the point of view of Database Systems (DBS), presenting the major reasons for the development of our DBS-based architecture of a KBMS. This architecture, called KRISYS, is then described, thereby concentrating on the mechanisms used to map its knowledge structures onto secondary storage as well as on the component responsable for the exploitation of the application access locality to increase performance.

2. DBS-based knowledge management

In looking for systems to undertake knowledge management tasks, one would directly arrive at the idea of applying DBMS for this purpose. Surely, when considering only knowledge maintenance, one may argue that existing DBMS are ideal for this task. The deficiencies found in KS for real world applications correspond exactly to the DBMS features.

2.1 Knowledge maintenance aspects

DBS-based knowledge maintenance means the cooperation of a KS for knowledge manipulation and a DBS for massive data management. In this context, interactions between KS and DBS can take place at any moment, when queries can be dynamically generated and transmitted to the DBS, and corresponding answers can be received and transformed into the KS internal representation. The DBS acts as a server to the KS, supplying on demand the data that the latter requires. The key issue is therefore the interface that allows the two systems to communicate. In developing such an interface, there are essentially two possibilities, depending on the moment when data requirements are transmitted: 'compiled' or 'interpretative' [MMJ84].

The compiled variation [CMT82,KY82,VCJ84] is based on two distinct phases (deduction and evaluation) which may be repeatedly performed. Firstly, a deduction on the side of the KS is executed by applying rules until a query containing no virtual relations can be sent for evaluation to the DBS. In other words, the evaluation is delayed until the query has been transformed into a conjunction of expressions involving only base relations which may now be translated to a DB query. The execution of the query on the side of the DBS and the subsequent delivery of the result to the former are then performed.

In the interpretative variation, deduction and evaluation phases are intermixed; that is, rule execution is interrupted as soon as a basis relation is needed. The corresponding query is then sent for evaluation to the DBS, which returns the result to the KS, allowing for continuation of rule execution [MMJ84].

Both the compiled and the interpretative variations have advantages and drawbacks. Commonly, the compiled approach generates fewer DBS requests than the interpretative strategy and DBS queries that are better to optimize. On the other hand, KS using the interpretative variation may execute their deductions without modifying their overall strategy. Finally, the compiled approach exhibits some problems in the evaluation of recursive virtual relations, and the interpretative one has limitations in the size of the retrieved data that must be maintained by the KS internally as long as a line of reasoning has not been completed.

In order to investigate the behavior of KS running in such a DBS-based KB management environment, we have constructed two prototype systems using the compiled and the interpretative approach. In the first system (DBPROLOG), a 'deductive' DBS has been implemented by means of the compiled variation on the basis of PROLOG [Ba87]. The other prototype, called DBMED, has applied the interpretative strategy to support a diagnosis XPS shell [HMP87,St86,Th87].

2.1.1 DBPROLOG

Overview

DBPROLOG uses the compiled approach implemented according to the meta-language level as described in [KY82,VCJ84] in order to take advantage of the query optimization mechanisms of its underlying DBS INGRES [SWKH76,Da87]. As a consequence, DBPROLOG cannot make use of the full expressive power of horn clauses. It only allows those clauses free of functions as well as those without arithmetical and logical expressions. Functions are not permitted because they cannot always be transformed into valid DB query statements, and logical and arithmetical expressions because they may often provoke an infinitely long search for solution [Za86].

It is important to mention here that these limitations occur in every coupling approach using a purely compiled strategy. Actually, the impossibility of evaluating recursive rules is also a drawback presented by most existing implementations. However, we have extended the compiled approach in order to deal with any type of recursive rules. To achieve this goal, the meta-language level of the compiled approach was enlarged with algorithms to recognize recursive predicates and strategies to evaluate them.

Recursion is evaluated by using an efficient least fixpoint iteration algorithm [Ba85], that generates just one DB query for each recursion step. Therefore, during the deduction process, recursive predicates are recognized by the meta-interpreter and removed from the deduction tree. The corresponding DB queries (following the applied algorithm for recursion handling) are evaluated and their results are stored in temporary relations. After this, the recursive predicate is introduced in the deduction tree once more, however, this time representing a base relation. The deduction process may then continue until a query containing only base relations can be sent for evaluation to INGRES.

A complete overview of a deduction process is shown in figure 1, where the numbers indicate the sequence of the individual steps taken in the process. For a complete description of the system and details of its implementation we refer to [Ba87].

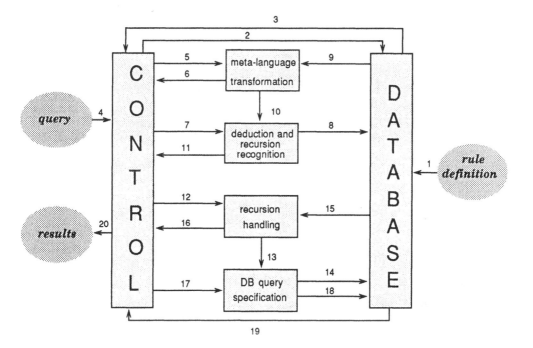

Figure 1: Overview of the deduction process of DBPROLOG [Ma89]

Investigation results

A detailed analysis of the behavior of the DBPROLOG prototype as well as many different performance measurements showed that inferences can be performed more efficiently by the DBS
component (figure 2). This turns out to be particularly evident as the amount of manipulated data
becomes larger. In spite of this fact, in every coupling environment, KS inference mechanisms
continue deciding when and how to access the DBS, leaving the latter working in a "blind" way.
This illustrates the performance bottleneck existing in any coupling approach due to the slave role
played by DBS in such a situation. The solution to this problem is to achieve a kind of system integration where DBS can "understand" the processing method of the XPS. By moving more information to the DBS, the DBS receives a higher-level specification of search requirements and
above all knowledge about the purpose of such requirements so that it has much more scope for
optimization. (Similar results to the above ones were also found by [Sm84]).

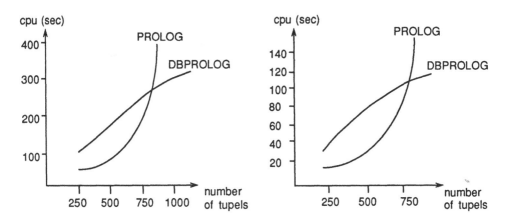

Figure 2: Measurement results of DBPROLOG: deduction in XPS vs. deduction in DBS

2.1.2 DBMED

Overview

Following this integration idea, we built the other coupling prototype. DBMED is a DBS-based
XPS shell prototype composed of

- the relational DBS INGRES [SWKH76,Da87] for knowledge maintenance, and

- the XPS shell MED1 [Pu83] for knowledge manipulation.

Starting from a simple coupling between MED1 and INGRES, we successively exploited the information about the XPS processing method in order to construct architectures that make use of DBS capabilities more intelligently (figure 3).

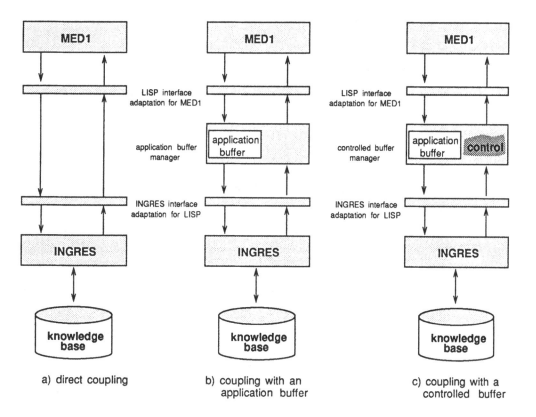

a) direct coupling b) coupling with an c) coupling with a
     application buffer controlled buffer

Figure 3: MED1/INGRES coupling overall architecture

Direct coupling

We started with a direct coupling between MED1 and INGRES implemented on the basis of the interpretative approach (figure 3a). Our goal in this first stage was to observe the access behavior of XPS and above all to investigate the adequacy of the coupling approach. For this reason, an important aim was to keep both XPS and DBS unchanged. The interface between MED1 and INGRES turned out, therefore, to be the existing interface between the KB and the problem solving component so that the corresponding operations could be mapped 1:1 to QUEL statements. This was achieved in two special modules as shown in figure 3a.

The investigation results of this coupling approach showed the inability of XPS working methods to process KB on secondary storage, emerged when consulting MED1 in this new environment.

Such an inability, which is described in [HMP87,St86,Th87], may be characterized here in summary by the slowing down of the system by a factor of 170 when compared to the main memory-based version of MED1 (table 1). In this coupling environment, the DBS works in a slave mode without knowledge about the XPS objects, about their relationships, as well as about the accesses of XPS so that it cannot make use of its optimization mechanisms.

Coupling with aplication buffer

The results of our first DBMED version showed that direct coupling was not sufficient for an effective XPS support. For this reason, an optimization of the system should occur by a kind of integration where the DBS capabilities could be better exploited.

This was achieved by the introduction of an application buffer between XPS and DBS which provides for the exploitation of the locality of the XPS accesses (figure 3b). Furthermore, the buffer enables the reduction of DBS calls as well as the reduction of the path length when accessing the objects of the KB. Additionally, the granules of the DB accesses could be enlarged (whole objects instead of attribute values) without provoking changes in MED1. A last, very important measure for optimization, was the complete delegation of the KB initialization (the first phase of a MED1 consultation) to the application buffer module. Thereby, the number of DB accesses was drastically reduced (from 2347 to 87) since the application buffer module applies few but adequate DB statements instead of initializing each attribute of each object sequentially.

The results of the above optimization measures can be characterized by the new performance of the system, i.e., a slowing down by only a factor of 18 which implies an improvement of a factor of 10.

Coupling with controlled buffer

Further optimizations were possible by an increased integration of both systems so that the processing characteristics of the XPS could be exploited for performance improvement. It was observed that in each phase of the problem solving process, XPS typically need different parts (i.e., objects) of the KB. These objects represent in reality the knowledge needed to infer the specific goal of the phase. For this reason, we say that these objects together build a context. (In [MW84], Missikoff and Wiederhold come to similar conclusions by observing the existence of problem and context related clusters). The optimization measure taken was therefore to make use of this knowledge in order to manage the application buffer. The idea was to fetch the required context during changes of processing phases causing an enormous reduction in subsequent accesses to the DB as well as in transfer overhead. Additionally, this approach allows for a better DBS access optimization due to the set-oriented specification of the needed objects. To achieve this goal, it was however necessary that MED1 looked ahead of its work, informing the buffer manager about the need for new contexts. The buffer manager knew about XPS working units so that it could fetch the corresponding object sets into the application buffer. It should be therefore clear that such a tight cooperation is only possible in a highly integrated environment.

Finally, our investigations have shown that the access frequencies of attributes of KB objects differ very much. Dynamic attributes (i.e., the ones that represent the knowledge inferred during a consultation or specific information about the case being analyzed) are accessed with very high frequencies; static attributes (i.e., the ones that represent the expert knowledge of the KS) on the other hand with very low frequencies. This is a consequence of the way reasoning mechanisms of KS use knowledge. Typically, each "piece-of" static knowledge, for example the action of a rule, is used just once during the reasoning process as the KS needs to infer a particular dynamic knowledge, i.e., the dynamic knowledge that this rule infers. However, each "piece-of" dynamic knowledge, for example the information that the patient has fever, is needed on several occasions. Since it is applied to infer several other "pieces-of" dynamic knowledge, i.e., the fever information can occur in the conditions of several rules, it is used and therefore accessed very frequently. Based on this observation, this coupling approach has splitted the static from the dynamic knowledge in order to make a more effective use of the controlled buffer.

The measurements of DBMED in this integrated fashion (i.e., with a controlled application buffer) showed a significant improvement of system performance (by a factor of 4.5) with regard to the previous version (see table 1).

	MED1 references	accesses to the KB (on sec. storage)	duration of a consultation in CPU-sec.	slow down factor with regard to main memory MED1
main memory based MED1	3574	3574	164	1
MED1 + INGRES (Coupling)	3574	3574	28105	~ 170
MED1 + INGRES + application buffer	1314	572	2946	~ 18
MED1 + INGRES + controlled application buffer	1314	93	643	~ 4

Table 1: Overview of measurement results of the DBMED prototype variations [Ma89]

2.2 Knowledge modeling and manipulation aspects

Existing DBS are adequate to satisfy knowledge maintenance requirements but are unable to support knowledge modeling and manipulation in an appropriate manner. The deficiencies of this support may be summarized as follows [Ma86a]:

- DBS are inadequate for an incremental KS development environment.

- Classical data models are not appropriate for representing knowledge. They lack expressiveness, semantics, and organizational principles, impose a strict distinction between object types and object instances, and exhibit very weak object-orientation [La83].

- DBMS provide insufficient manipulation support. They have only very limited reasoning capabilities.

- Integrity control mechanisms are too weak. DBMS cannot guarantee an automatic maintenance of complex semantic integrity constraints.

NDBS approach

As a matter of fact, existing DBMS are not only unable to support KS. In reality, they also fail to meet the requirements found in a range of other emerging application areas such as CAD/CAM, geographic information management, VLSI design, etc. The inherent causes for this situation were illustrated already at the beginning of the 1980's [Ea80,Si80,Lo81,GP83,Lo83,HR85,Lo85] being pointedly characterized in the hypothesis proposed in [HR83]:

"The general purpose DBMS are general only for the purpose of commercial applications, and their data models are general only for modelling commercial views of the world."

These DBS deficiencies can be roughly expressed as "the lack of an orientation towards the application requirements". This observation led to the development of so-called Non-Standard DBS and the corresponding NDBS kernel architecture [HR85,Da86,PSSWD87]. In such an architecture (figure 4), the required orientation is achieved on the top of the supported data model interface in an additional NDBS component, called application layer (AL).

The basic idea behind this approach is to keep a strong separation between kernel and AL so that the kernel remains application-independent. That is, it efficiently realizes neutral, yet powerful mechanisms for management support of all types of non-standard application: storage techniques for a variety of object sizes, flexible representation and access techniques, basic integrity features, locking and recovery mechanisms, etc. In other words, the kernel can be viewed as the implementation of a non-standard data model which, in turn, characterizes its external interface. On top of this interface, there is an application-oriented AL, which is responsible for the realization of mechanisms useful for a specific application class. In our case, such an AL should support the above mentioned points (i.e., knowledge modeling and manipulation tasks). Thus, the interface supported by the AL is at a much higher level than the data model interface. Its orientation is more towards the purpose of its application class, whereas the data model interface offers a more or less neutral support corresponding to the tasks involved with knowledge maintenance.

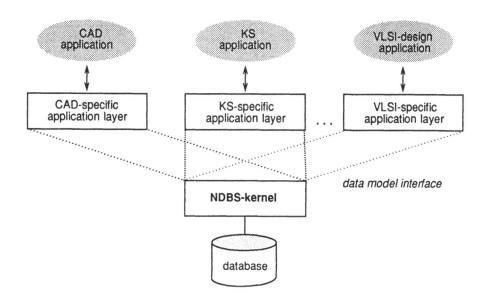

Figure 4: Overall architecture of NDBS

Summarizing these ideas, we argue that NDBS are adequate for knowledge management. NDBS kernels and corresponding data models are much richer than conventional DBS and should, for this reason, achieve maintenance tasks in a more efficient and effective manner. Furthermore, since knowledge modeling and manipulation tasks are usually achieved at a higher level than knowledge maintenance tasks, NDBS architectural conception seems to be quite advantageous.

A practical investigation

With the purpose of validating our premises about the appropriateness of the architectural concept of NDBS, we have investigated the applicability of a non-standard data model as a neutral instrument for mapping application-oriented structures and operations at the kernel interface. For this investigation, we have chosen a simple frame-based representation as an example of an orientation towards KS. Here, we just summarize our results (figure 5), referring to [Ma86b, HMM87,Mi88b] for a detailed description of the performed analysis.

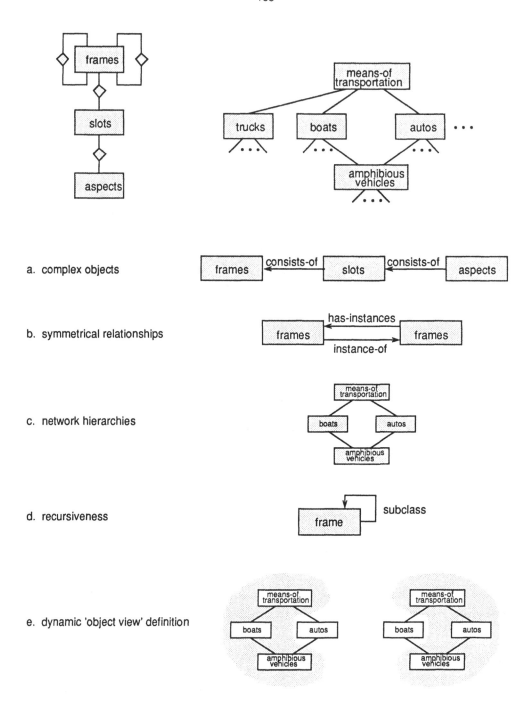

Figure 5:   Frame-based representation schemes from a structural point of view

First, one may observe the occurrence of only three different constructs in frame-based representation schemes: *frames* for the representation of real world objects, *slots* representing attributes, and *aspects* for further descriptions of the slots (e.g., by means of integrity constraints). From a structural point of view, frames are therefore composed of slots, which, in turn, consist of aspects, suggesting the DB schema of *complex objects*, i.e., those whose internal structures (the components) are also objects of the DB (figure 5a). Second, one may note that frames in general possess predefined slots expressing organizational axes (such as instance-of for classification and subclass-of for generalization), which represent *symmetric n:m relationships* (figure 5b) between frames (back references are respectively has-instances and has-subclasses). These relationships are specially important because they provide different ways to organize frames into *network hierarchies* (figure 5c). For the construction of such hierarchies, one has to *recursively* follow a particular relationship in order to structure frames of a higher level (figure 5d). Finally, one may observe that during the manipulation of a frame one is defining a kind of *dynamic view* of this frame since one is always dealing with only part of the frame hierarchy (in figure 5e, one is 'viewing' the properties of amphibious vehicles considering their characteristics of either an auto or a boat).

Summarizing, one may conclude that a data model for mapping frame-based representation schemes must provide means for the support of complex objects, of a direct and symmetric representation of n:m relationships and, consequently, network hierarchies involving recursiveness, and of a dynamic 'object view' definition. Since these concepts are incorporated in most non-standard data models, which support the so-called structural object-orientation [Di86,Di87], one may argue that the generic constructs offered by these models allow for an accurate and efficient mapping of the KS-oriented structures. The main philosophy behind these models is to support a precise representation as well as a dynamic construction of complex objects (here frames), using elementary building blocks that are also represented as objects of the DB (here slots and aspects). In this sense, a direct and symmetric representation of relationships (including n:m) and recursiveness play a very important role in this context. (For further comments about the adequacy of non-standard data models (specially MAD) for KS see [Mi88b,HMM87]).

3. KRISYS Approach

Reflecting the aspects of the NDBS approach and based on the described investigation results, a DBS-based architecture for KB management was developed [Ma89, Ma88b]. This architecture, called KRISYS, is architecturally divided in two major modules (see figure 6).

On the top of a NDBS kernel, three different components (corresponding to the AL of a NDBS)

- implement the application interface of KRISYS by means of a language that provides the user or KS with an abstract, functional view of the KB [DLM90],

- provide a mixed knowledge representation framework supporting object-centered representation, abstraction concepts, reasoning facilities (rules), attached procedures (demons), and object-orientation by means of methods [MM89], and

- embody the "nearby application locality" concept, thereby reducing DBS calls as well as the path length when accessing the objects of the KB [LM89].

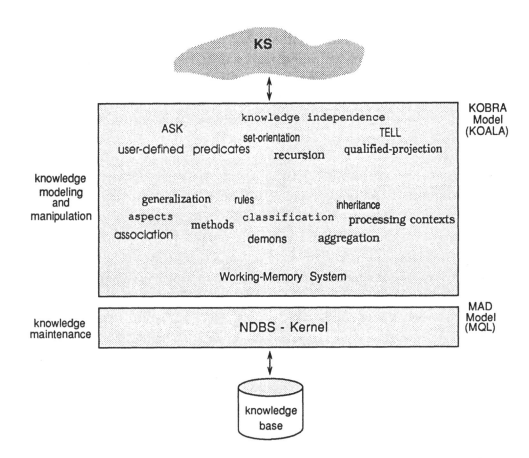

Figure 6: Overview of the KRISYS architecture

The use of the NDBS architectural concept is quite advantageous in the KRISYS architecture. Knowledge modeling and manipulation tasks, which are in general achieved at a higher level than knowledge maintenance tasks, are supported by the upper components of the system, whereas the management of the KB is undertaken by the NDBS kernel component. Following this issue, the semantics of our knowledge structures remains outside of the NDBS kernel which views them simply as a kind of complex objects to be adequately managed.

The DBMS kernel chosen for KRISYS, named PRIMA [Hä88a, HMMS87], was developed for the support of non-standard applications. PRIMA, PRototype Implementation of the MAD model [Mi88a, Mi89a], provides dynamic definition and handling of complex objects based on direct and symmetric management of network structures and recursiveness enabling the mapping of our knowledge structures in an effective and straightforward manner.

3.1 Working-Memory System

Architecture

The nearby application locality concept mentioned above is achieved by a component, denoted Working-Memory System (WMS), that realizes a kind of application buffer. It temporarily stores needed objects in a main-memory structure, called working-memory, that offers almost direct access at costs comparable to a pointer-like access. WMS supports a processing model aimed at a high locality of object references, thereby drastically reducing the path length when accessing the KB. To reach this end, WMS exploits the concept of processing contexts as described in section 2.1.2. This concept is realized by the context manager, which fetches and discards such contexts as notified by specific control calls. These calls are then transformed into set-oriented kernel operations (complex queries) to extract the specified object from the DB or to discard them from the working-memory. Access requests to KB objects or parts thereof are dealt with by the working-memory manager, which generates and sends simple queries to the kernel if the requested objects are not found in the working-memory. The distribution component is only responsible for proper routing of the respective calls.

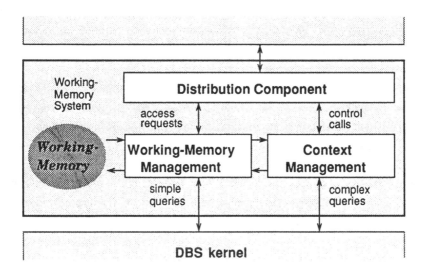

Figure 7: Architecture of the Working-Memory System [Ma89]

Performance Comparison

The performance of our working-memory component has been compared with the performance of the investigated coupling approaches between KS and DBS as shown in table 2. To pursue this comparative investigation, we have adapted the diagnosis XPS shell MED2 [Pu86a,Mi89b] to work in different environments as shown in the table. During our consultations, MED2 has exploited a 7MB KB to diagnose fungi. Hence, we have repeatedly performed an identical application with the same performance critical parameters in each of the coupling/integration approaches. Therefore, our measurement results seem to be indicative for the quality of these investigated approaches.

	MED2 references	accesses to the KB (on secondary storage)
MED2 + DBS (direct coupling)	21303	21303
MED2 + DBS + application buffer	21303	8427
MED2 + KRISYS	23437	1105

Table 2: Overview of measurement results during a typical consultation [Ma89]

Coupling DBS and KS directly is obviously the most inefficient alternative. As already mentioned, every KS reference to a KB object has to be translated into a DBS call. This approach not only generates long execution paths (from application to DB-buffer) but may also provoke many accesses to the KB, depending on the DB-buffer size and on the required working sets.

To reduce the long path lengths, an application buffer may be applied in order to keep the most recently used objects close to the KS. However, since this approach does not make use of knowledge about the KS access behavior, a large number of DBS calls is still necessary especially after changes of processing phases when most of the requested objects are not found in the buffer.

KRISYS eliminates this problem by exploiting knowledge about the KS processing model. Since the replacement of KB objects in the working-memory is carried out in accordance with the processing contexts of the KS, DBS calls as well as accesses to secondary storage and transfer overhead are strongly minimized. (Note that the number of calls increases because of the XPS control calls to inform KRISYS about the course of the problem solving process).

3.2 Mapping knowledge structures

On the basis of the mapping mechanism implemented in KRISYS, we started to develop some knowledge-based applications in order to validate our new concepts and research ideas. In the mean time, there are about 10 applications running on KRISYS, including expert systems for architectural design or fungi diagnosis as well as some applications from the CAD/CAM field.

In a KBMS environment, the way such applications are developed turns out to be quite different. In contrast to existing DB techniques, KRISYS advocates a modeling process in a stepwise fashion (figure 8). That is, in each step, only some of the details of the problem are considered so that those details less relevant to the subproblem at hand are neglected until some later step [MM89,BMW84]. In such environments, the modeler takes advantage of the reasoning facilities provided by the abstraction concepts [Ma88a,RHMD87] in order to facilitate his task by exploiting the system to make deductions about the objects as well as to guarantee the structural and semantic integrity of the KB.

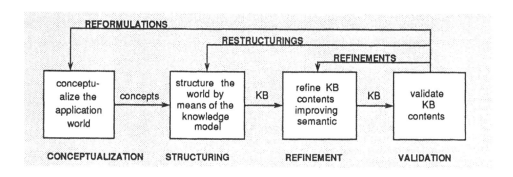

Figure 8: Evolutionary process of KS development [MM89]

So, in KBMS environments the first step in developing a KS is to conceptualize the real world situation of its application domain (conceptualization phase). The result is expressed by means of a knowledge model (structuring phase) which is refined in a stepwise manner in order to improve it with more semantics (refinement). A validation phase is then performed causing a kind of feedback to the previous ones in order to enrich as well as to correct the KB incrementally until the KS exhibits the same level of know-how as human experts.

An investigation of these applications has shown that during this modeling process, accesses to the KB are mostly caused by functions that provoke modifications on the internal structures of the KB (i.e., the KB schema). For this reason, the mapping mechanism should flexibly support functions that modify the internal KB structures (figure 9a). However, during a KS consultation, accesses to the KB are mostly restricted to read and write operations (i.e., manipulation of single attributes or instances of some object type). Insert and delete operations may also occur, but only at the level of instances. In other words, during a consultation, modifications in the structures of the KB (e.g., changing object types, adding new types or integrity constraints) either do not occur or are very seldom. Consequently, the chosen mapping mechanism should give more priority to the optimization of functions that do not modify the structures of the KB (figure 9b).

a) during the modeling process:

- structuring — -definition of objects, attributes, relationships
- refinement — -definition of aspects, demons, rules
- validation — -simulation of the application
- restructuring — -deletion of objects
  -redefinition of attributes, relationships
- refinements — -deletion of aspects, demons, rules
  -redefinition of aspects

↳ functions that provoke modifications on the KB schema

b) during the operation of an application:

- operation — -access to KB objects
  -search on object hierarchies
  -definition and manipulation of instances
  -message passing

↳ functions that do not provoke modifications on the KB schema

Figure 9: Overview of the kind of accesses to the KB during the life time of a KS

Since the efficiency of KRISYS functions directly depends on the chosen DB schema together with the corresponding mapping mechanism, the necessity of providing two different DB schemas became then apparent: one for the construction or modeling phase, that efficiently supports functions that modify KB structures, and another for the operation phase that will exploit the strong KB stability to give high priority to the optimization of functions for storage and retrieval.

Another very important observation was that after the modeling process, the structures of the KB of the KRISYS applications differ very much (figure 10). They differ, for example, by means of the number, kind, height, and form of abstraction hierarchies specified, of the distribution of instances over these hierarchies and per class, of the frequency of attributes (re)definitions as well as specializations of integrity constraints, etc. Therefore, the necessity of providing an application-oriented DB schema with the corresponding mapping mechanism for the operation phase of each KRISYS application became also apparent.

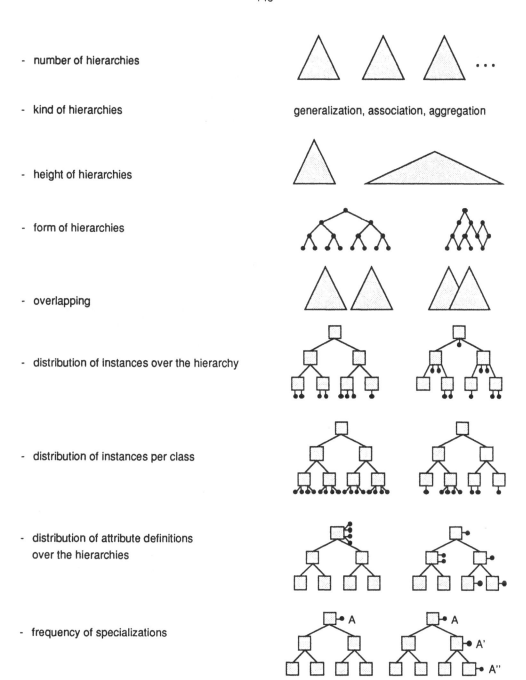

- number of hierarchies

- kind of hierarchies

  generalization, association, aggregation

- height of hierarchies

- form of hierarchies

- overlapping

- distribution of instances over the hierarchy

- distribution of instances per class

- distribution of attribute definitions
  over the hierarchies

- frequency of specializations

- kind of operational knowledge

  methods, demons, rules

- definiton of processing contexts

Figure 10: Some differences of KB-structures of KRISYS applications

The idea is then to extend the development process of KS with a latter phase, called optimization, during which KRISYS automatically performs a DB schema transformation in order to generate an application-oriented, and consequently very effective, DB schema for this KS (figure 11). The most appropriate DB schema for a KS, which depends on characteristics of the application, is then determined by means of an analysis of the structures of the KB as well as of the processing contexts of the KS, that are completely specified and explicitly represented in KRISYS just before the optimization takes place [MM88,LM89].

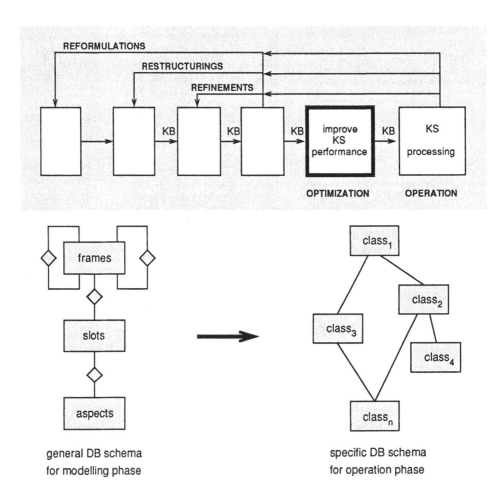

general DB schema
for modelling phase

specific DB schema
for operation phase

Figure 11: Evolutionary process of KS development: the optimization phase

Once the new DB schema is generated, a compilation of the KB is then performed, thereby also precompiling the application queries encoded within methods, demons, and rules. In other words, the optimization phase represents the "exit" of an interpretative (flexible but probably inefficient) KB manipulation, which is necessary during KB construction, and the "entry" into a compiled (and therefore efficient) KB manipulation, which is necessary during KS operation.

4. Summary

Systems for KB management should integrate features for supporting knowledge modeling, manipulation, and maintenance. For this reason, knowledge management requests a combination of AI and DBS technologies, demanding, however, much more than just coupling DBS as a back-end storage system to KS. To achieve the best functionality and performance, these technologies must be integrated in order to allow for a favorable combination of efforts rather than for the work of one against the other.

To achieve such integration, *a new generation of systems, so-called KBMS, specially constructed for an effective and efficient management of knowledge bases* must be designed. In developing KBMS, NDBS-based architectures seem to be very appropriate since they provide effective means for such an integration: the NDBS kernel fulfils knowledge maintenance, and the application layer supports knowledge modeling and manipulation.

KRISYS is a KBMS prototype constructed according to this NDBS architectural concept. It supports a rich spectrum of concepts for knowledge modeling and manipulation, including object-orientation, data-driven computation, reasoning mechanisms, abstraction concepts, and a rich knowledge language. If offers an efficient framework for the exploitation of the application locality, taking advantage of the knowledge about the access behavior of the KS in order to

- make use of the DBMS kernel mechanisms to improve access efficiency to secondary storage,
- reduce the path length when accessing KB objects, and
- minimize I/O operations as well as transfer overhead.

KRISYS accomplishes therefore fast access to stored KB objects by exploiting the existence of processing contexts whose contents are kept close to the application in its working-memory.

Finally, it provides an effective and efficient support of the modeling process and of the application processing by means of different mapping mechanisms to the NDBS kernel: a general mechanism is provided during KS construction in order to allow for a flexible KB manipulation, and a specific, application-oriented in order to guarantee efficient KS operation.

References

Ba85   Bayer, R.: Database Technology for Expert Systems, in: Proc. GI-Conference "Knowledge-based Systems", IFB 112, Springer-Verlag, Munich, Oct. 1985, pp. 1-16.

Ba87   Bauer, S.: A PROLOG-based Deductive Database System (in German), Undergraduation Final Work, University of Kaiserslautern, Computer Science Department, Kaiserslautern, 1987.

BM86   Brodie, M.L., Mylopoulos, J. (eds.): On Knowledge Base Management Systems (Integrating Artificial Intelligence and Database Technologies), Topics in Information Systems, Springer-Verlag, New York, 1986.

BMS84   Brodie, M.L., Mylopoulos, J., Schmidt, J.W. (eds.): On Conceptual Modelling (Perspectives from Artificial Intelligence, Databases, and Programming Languages), Topics in Information Systems, Springer-Verlag, New York, 1984.

BMW84   Borgida, A., Mylopoulos, J., Wong, H.K.T.: Generalization/Specialization as a Basis for Software Specification, in: [BMS84], pp. 87-114.

CMT82   Chakravarthy, U.S., Minker, J., Tran, D.: Interfacing Predicate Logic Languages and Relational Data Bases, in: Proc. of the 1st Int. Logic Programming Conf., Marseille, Sept. 1982.

Da83   Date, C.J.: An Introduction to Database Systems, Vol. 1 and Vol. 2, 3rd edition, Addison-Wesley Publishing Company, Reading, Mass., 1983.

Da86   Dadam, P., et al.: A DBMS Prototype to Support Extended NF^2-Relations: An Integrated View on Flat Tables and Hierarchies, in: Proc. ACM SIGMOD Conf., Washington, D.C., 1986, pp. 356-367.

Da87   Date, C.J.: A Guide to INGRES, Addision-Wesley Publishing Company, Reading, Mass., 1987.

DLM90   Deßloch, S., Leick, F.J., Mattos, N.M.: An Approach to Knowledge Base Languages, Research Report, University of Kaiserslautern, 1990, submitted for publication.

Ea80   Eastman, C.M.: System Facilities for CAD-Databases, in: Proc. 17th Design Automation Conf., Minneapolis, 1980, pp. 50-56.

GP83   Gründig, L., Pistor, P.: Land Information Systems and Their Requirements on Database Interfaces (in German), IFB 72, Springer-Verlag, 1983, pp. 61-75.

Hä88a   Härder, T. (ed.): The PRIMA Project Design and Implementation of a Non-Standard Database System, SFB 124 Research Report No. 26/88, University of Kaiserslautern, Kaiserslautern, 1988.

HMM87   Härder, T., Mattos, N., Mitschang, B.: Mapping Frames with New Data Models (in German), in: Proc. German Workshop on Artificial Intelligence GWAI'87, Springer Verlag, Geseke, 1987, pp. 396-405.

HMMS87   Härder, T., Meyer-Wegener, K., Mitschang, B., Sikeler, A.: PRIMA - A DBMS Prototype Supporting Engineering Applications, SFB 124 Research Report No. 22/87, University of Kaiserslautern, 1987; in: Proc. 13th VLDB Conf., Brighton, UK, 1987, pp. 433-442.

HMP87   Härder, T., Mattos, N.M., Puppe, F.: On Coupling Database and Expert Systems (in German), in: State of the Art, Vol. 1, No. 3, pp. 23-34.

HR83   Härder, T., Reuter, A.: Concepts for Implementing a Centralized Database Management System, in: Proc. of the International Computing Symposium 1983 on Application Systems Development, Nürnberg, Teubner-Verlag, March 1983, pp. 28-59.

HR85   Härder, T., Reuter, A.: Architecture of Database Systems for Non-Standard Applications (in German), in: Proc. GI-Conference on Database Systems for Office, Engineering and Science Environments, IFB 94, Springer-Verlag, Karlsruhe, 1985, pp. 253-286.

Ke86   Kerschberg, L. (editor): Proceedings from the First International Workshop on Expert Database Systems, Kiawah Island, South Carolina, October 1984, Benjamin/Cunnings Publ. Comp., Menlo Park,CA., 1986.

KY82   Kunifuji, S., Yokota, H.: PROLOG and Relational Data Bases for Fifth Generation Computer Systems, in: Workshop on Logical Bases for Data Bases, Toulouse, Dec. 1982.

La83   Lafue, G.M.E.: Basic Decisions about Linking an Expert System with a DBMS: A Case Study, in: Database Engineering, Vol. 6, No. 4, Dec. 1983, pp. 56-64.

Lo81   Lorie, R.A.: Issues in Databases for Design Applications, in: Proc. IFIP Conf. on CAD Data Bases, File Structures and Data Bases for CAD, (eds.: Encarnacao, J., Krause, F.L.), North-Holland Publ. Comp., 1981, pp. 214-222.

Lo83   Lohman, G., et al.: Remotely-Sensed Geophysical Databases: Experience and Implications for Generalized DBMS, in: Proc. of the SIGMOD'83 Conference, San Jose, 1983, pp. 146-160.

Lo85   Lockemann, P.C.; et al.: Requirements of Technical Applications on Database Systems (in German), in: Proc. GI-Conference on Database Systems for Office, Engineering and Science Environments, IFB 94, Springer-Verlag Karlsruhe, 1985, pp. 1-26.

LM89   Leick, F.J., Mattos, N.M.: A Framework for an Efficient Processing of Knowledge Bases on Secondary Storage, in: Proc. of the 4th Brazilian Symposium on Data Bases, Campinas-Brazil, April 1989.

LS87   Lockemann, P.C., Schmidt, J.W. (eds.): Database-Handbook (in German), Springer-Verlag, Berlin, 1987.

Ma86a   Mattos, N.M.: Concepts for Expert Systems and Database Systems Integration (in German), Research Report No. 162/86, University of Kaiserslautern, Computer Science Department, Kaiserslautern, 1986.

Ma86b   Mattos, N.M.: Mapping Frames with the MAD model (in German), Research Report No. 164/86, University of Kaiserslautern, Computer Science Department, Kaiserslautern, 1986.

Ma88a   Mattos, N.M.: Abstraction Concepts: the Basis for Data and Knowledge Modeling, ZRI Research Report No. 3/88, University of Kaiserslautern, in: 7th Int. Conf. on Entity-Relationship Approach, Rom, Italy, Nov. 1988, pp. 331-350.

Ma88b   Mattos, N.M.: KRISYS - A Multi-Layered Prototype KBMS Supporting Knowledge Independence, ZRI Research Report No. 2/88, University of Kaiserslautern, in: Proc. Int. Computer Science Conference - Artificial Intelligence: Theory and Application, Hong Kong, Dec. 1988, pp. 31-38.

Ma89   Mattos, N.M.: An Approach to Knowledge Base Management - requirements, knowledge representation and design issues -, Doctoral Thesis, University of Kaiserslautern, Computer Science Department, Kaiserslautern, 1989.

Mi88a   Mitschang, B.: A Molecule-Atom Data Model for Non-Standard Applications - Requirements, Data model Design, and Implemlentation Concepts (in German), Doctoral Thesis, University of Kaiserslautern, Computer Science Department, Kaiserslautern, 1988.

Mi88b   Mitschang, B.: Towards a Unified View of Design Data and Knowledge Representation, in: Proc. of the 2nd Int. Conf. on Expert Database Systems, Tysons Corner, Virginia, April 1988, pp. 33-49.

Mi89a   Mitshang, B.: Extending the Relational Algebra to Capture Complex Objects, in: Proc. of the 15th VLDB Conf., Amsterdam, 1989, pp. 297-306.

Mi89b   Michels, M.: The KBMS KRISYS from the Viewpoint of Diagnosis Expert Systems (in German), Undergraduation Final Work, University of Kaiserslautern, Computer Science Department, Kaiserslautern, 1989.

MM88 Mattos, N.M., Michels, M.: Modeling Knowledge with KRISYS: the Design Process of Knowledge-Based Systems Reviewed, Research Report, University of Kaiserslautern, 1988.

MM89 Mattos, N.M., Michels, M.: Modeling with KRISYS: the Design Process of DB Applications Reviewed, in: Proc. the 8th Int. Conf. on Entity-Relationship Approach, Toronto - Canada, Oct. 1989, pp. 159-173.

MMJ84 Marque-Pucheu, G., Martin-Gallausiaux, J., Jomier, G.: Interfacing PROLOG and Relational Data Base Management Systems, in: New Applications of Data Bases (eds. Gardarin, G., Gelenbe, E.), Academic Press, London, 1984, pp. 225-244.

MW84 Missikoff, M., Wiederhold, G.: Towards a Unified Approach for Expert and Database Systems, in: [Ke86], pp. 383-399.

PSSWD87 Paul, H.-B., Schek, H.-J., Scholl, M.H., Weikum, G., Deppisch, U.: Architecture and Implementation of the Darmstadt Database Kernel System, in: ACM SIGMOD Conf., San Francisco, 1987, pp. 196-207.

Pu83 Puppe, F.: MED1: a Heuristic-based Diagnosis System with an Efficient Control Structure (in German), Research Report No. 71/83, University of Kaiserslautern, Computer Science Department, Kaiserslautern, 1983.

Pu86a Puppe, F.: Diagnostic Problem Solving with Expert Systems (in German), Doctoral Thesis, University of Kaiserslautern, Computer Science Department, Kaiserslautern, 1986.

Pu86b Puppe, F.: Expert Systems (in German), in: Informatik Spektrum, Vol. 9, No. 1, February 1986, pp. 1-13.

Pu88 Puppe, F.: Introduction in Expert Systems (in German), Springer-Verlag, Berlin, 1988.

RHMD87 Rosenthal, A., Heiler, S., Manola, F., Dayal, U.: Query Facilities for Part Hierarchies: Graph Traversal, Spatial Data, and Knowledge-Based Detail Supression, Research Report, CCA, Cambridge, MA, 1987.

Si80 Sidle, T.W.: Weakness of Commercial Data Base Management Systems in Engineering Application, in: Proc. 17th Design Automation Conf., Minneapolis, 1980, pp. 57-61.

Sm84 Smith, J.M.: Expert Database Systems: A Database Perspective, in: [Ke86] , pp. 3-15.

St82   Stefik, M.J., et al.: The Organization of Expert Systems, A Tutorial, in: Artificial Intelligence, Vol. 18, 1982, pp. 135-173.

St86   Stauffer, R.: Database Support Concepts for the Expert System MED1 (in German), Undergraduation Final Work, University of Kaiserslautern, Computer Science Department, Kaiserslautern, 1986.

ST89   Schmidt, J.W., Thanos, C. (eds.): Foundations of Knowledge Base Management (Contributions from Logic, Databases, and Artificial Intelligence), Topics in Information Systems, Springer-Verlag, Berlin, 1989.

SWKH76   Stonebraker, M., Wong, E., Kreps, P., Held, G.: The Design and Implementation of INGRES, in: ACM TODS, Vol. 1, No. 3, 1976, pp. 189-222.

Th87   Thomczyk, C.: Concepts for Coupling Expert and Database Systems - an Analysis Based on the Diagnose Expert System MED1 and the Database System INGRES (in German), Undergraduation Work, University of Kaiserslautern, Computer Science Department, Kaiserslautern, 1987.

Ul82   Ullman, J.D.: Principles of Database Systems, 2nd edition, Computer Science Press, London, 1982.

VCJ84   Vassiliou, Y., Clifford, J., Jarke, M.: Access to Specific Declarative Knowledge by Expert Systems, in: Decision Support Systems, Vol. 1, No. 1, 1984.

Za86   Zaniolo, C.: Safety and Compilation of Non-recursive Horn Clauses, in: Proc. First Int. Conf. on Expert Database Systems, Charleston, South Carolina, April 1986, pp. 167-178

Knowledge Bases and Databases:
Current Trends and Future Directions

John Mylopoulos[1]
Michael Brodie[2]

"...companies that bought the first expert systems found themselves with segregated islands in the middle of their data processing environment..."

Harry Reinstein, Aion Corporation[3]

1. Introduction

1.1. Motivation

Building computer systems that *manage* and *interpret* large amounts of information is an old idea, as old as computer science. After a decade of practice in building expert systems, we are concluding that for any expert system technology to mature, somehow, this old idea will have to be realized. We will try to outline the reasons for requiring from a computer system both access to large amounts of information and the ability to interpret that information. We will also characterize the main approaches to a solution. In the process, we hope to convince the reader that deep solutions to the problem at hand lead to a new perspective on knowledge bases and knowledge representation systems which transcend both expert system and database technologies as we now know them.

Knowledge bases are not a novelty any more. They have been studied in research labs since the late '60s and have seen commercial applications since the beginning of the '80s through expert systems. The number of deployed expert systems was estimated to be 1,400 in 1988, up from 50 the year before.[4] Moreover, it is estimated that over 10,000 expert systems will be deployed or under development this year (1990).

So, after a decade of claims, trials and errors, successes and failures, knowledge bases seem to be coming of age! Their introduction into an industrial setting has led to several conclusions. Some of those conclusions focus on knowledge representation and reasoning features and their usefulness or adequacy in addressing "real world" problems. Others are more pragmatic and address issues of embedding knowledge bases into pre-existing conventional data processing environments. It is the latter that have raised a general interest in interfacing knowledge bases and databases.

Why interface knowledge bases and databases?

First, and most importantly, deployed expert systems often need to be embedded within larger, usually existing, data processing systems. The American Express credit verification system is a good example of an expert system requiring access to several databases. The system, intended to assist a human operator in verifying a client's credit, has reduced information presented to the human operator from 16 screens down to 1 thereby making 30 second

---

[1] Department of Computer Science, University of Toronto, 10 King's College Road, Toronto CANADA M5S 1A4.

[2] GTE Laboratories, 40 Sylvan Road, Waltham MA 02254.

[3] Reinstein's quote taken from an article titled "Artificial Intelligence is Starting to Get Smart," published in the Toronto Star, January 8, 1990.

[4] From Raj Reddy's AAAI presidential address '88, [REDD88]

decisions possible. The system uses KEE to access information from 6 conventional databases including IMS and DB2.

Experts looking for bottlenecks in the development of an expert system technology seem to agree on this point. See, for example, Reinstein's motivating quote above, but also those of others such as the following from d'Agapeyeff, chairman of ExperTech (UK expert system shell vendor) who states in a report to the Alvey directorate:

> *"...the scope and range of utilization of expert systems may depend on...the ability to access existing files or databases..."[1]*

For this embedding to be done, communication needs to be established at a hardware, systems and applications level between the two systems. Many early difficulties of interfacing AI hardware and systems software with conventional data processing hardware and software have been overcome. Unfortunately; achieving this embedding still is costly and error prone task often leading to inefficient or unacceptable solutions.

A related reason for wanting this interface is that direct access to a database is the best way to acquire and justify statistical knowledge used by an expert system. The application of inductive learning techniques to databases to acquire new knowledge is now attracting considerable attention in the research community. In this chapter, this means that some functions we are coming to expect from expert systems **demand** access to databases.[2]

A second important impetus for the interest in interfacing knowledge bases with databases is commercial and has to do with the market strategy of expert system shell (ESS) vendors. Consider the facts: the size of the market for database management systems (DBMSs) is estimated to be an order of magnitude greater than that for expert system building tools.[3] The number of databases deployed more than three orders of magnitude greater than the number of deployed expert systems[4]. From the perspective of an ESS vendor, it makes some sense to market the ESS as a tool that can be used to build on top of a deployed database than as a standalone technology. Indeed, it appears that this is part of the selling strategy of a new wave of expert system companies.[5]

A third, complementary reason for wanting compatible technologies for building knowledge bases and databases focuses on the problem of how does a corporation introduce an expert system group within its organizational structure. Making that group part of the corporation's information systems department seems like a sensible solution, if the expert system group is not perceived as a "segregated island in the middle of a data processing department," to borrow Reinstein's quote. Making knowledge base access to databases an integral part of expert

---

[1] From Ruth Kerry, "Integrating Expert Systems and Databases," Central Computer and Communications Agency, British Government, 1989.

[2] Statistical learning from databases has become an important knowledge acquisition technique for expert systems, receiving increasing attention in the last few years. See, for example, proceedings of AAAI and IJCAI workshops on the topic held in the last three years.

[3] The dollar value of DBMSs sold in 1987 was $2.2B and will reach $5.8B by 1991 [Datamation, June 16, 1989]. The dollar value of the expert system market was estimated to be $135M in 1989 and will rise to $200M in 1990 [Toronto Star, Janury 8, 1990].

[4] The number of deployed DBMSs was estimated to be 4.9M, including 600,000 mainframe-based and 4.3M PC-based.

[5] Again, quotes from the HiTech literature support this claim:

> *"... If expert system technology is to be admitted into the mainstream of American Business, the door through which it will pass might well be labelled 'DBMSs'... "*
> HighTech Business, 1988

system technology could even help in facilitating the introduction and acceptance of the technology within an organization.

In summary, interfacing knowledge bases and databases makes sense from the perspective of the expert system practitioner, but also the point of view of the expert system industry and even that of the corporate users who are struggling to find a home within their organizations for this new and not always conventional technology.

1.2. An Example: The General Motors Dealer Review Advisor[1]

Details from a case study can further motivate the need for interfacing knowledge bases with a conventional data processing environment. Consider ANALYST, developed by GM to aid its credit analysts in 230 GM Acceptance Corporation branch offices across North America responsible for granting inventory financing to GM dealers. Part of the credit analyst's job is to evaluate the risk of a particular credit decision by analyzing the dealer's past performance, local economy and operating ability. By this analysis, a prediction is made about the dealer's probable performance with a recommendation for a credit line.

The benefits envisioned by the deployment of such an expert system include faster reviews, reduced training for staff, consistency in decision making, greater compliance to company policy, reduced lending risks and more time for the analysts to spend on the difficult cases.

ANALYST interacts with the human credit analyst to help him or her come to a recommendation (Figure 1). In the process, ANALYST uses well over a thousand data elements during a review and allows the credit analyst to display several more thousands in forms, tables and graphs.

The centre piece of each analysis is the dealer's financial statement. Background information on the dealer is available from on-line databases (Dealership Database). Off-line feeds into ANALYST happen nightly or monthly (GM and GMAC mainframe applications and databases), depending on the source. Financial statement data not available from on-line sources are clerically entered.

ANALYST was deployed as a distributed system (Figure 2). Receipt, processing, storage and preparation for distribution of dealership data is done by batch jobs running on a mainframe (ANALYST support system). This system communicates and gathers information from several other mainframes. Individual ANALYST knowledge bases reside on workstations at branch offices. All communications between the workstations and the host mainframes occur over a conventional communications network.

It should be clear that for a system such as ANALYST, effective access to corporate databases by ANALYST knowledge bases is maybe the most critical issue in the operation of the system.

---

[1] Description of ANALYST based on Hutson, M. A. and Lamoree, S. P., "The Dealer Review Advisor," *Sun Technology*, Winter 1989, 78-84.

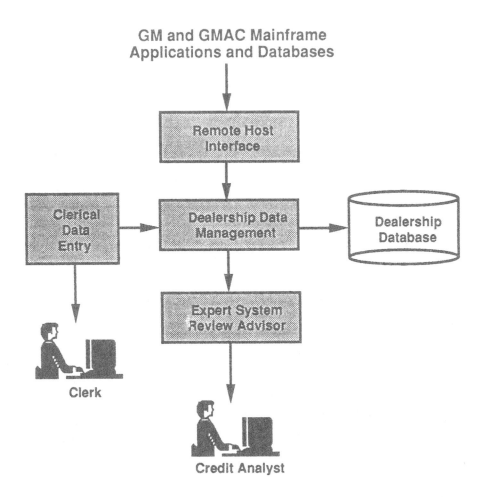

Figure 1

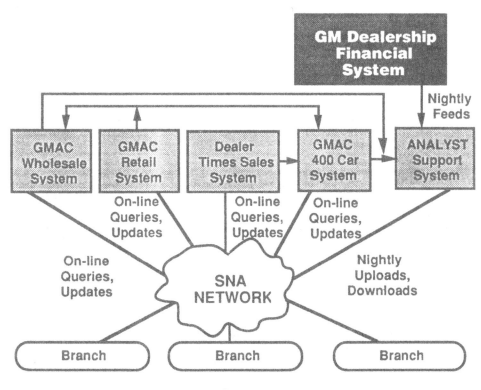

Figure 2

1.3. Terminology

Before proceeding with a review of issues, a few words about the terminology we'll be using in the sequel:

Database - a collection of persistent (stored) operational data used by the application systems of a particular enterprise.[1]

Database management system (hereafter **DBMS**) - a generic system for constructing and accessing databases. Its management facilities include means for ensuring database persistence, sharing, efficient access and error recovery.

Knowledge base - a data structure representing propositions about some application.[2] The language used for the propositions must come with a truth theory which prescribes what knowledge base operations are meaningful or desirable. An operation which selects every other symbol of the propositional interpretation of a knowledge base, for example, is clearly meaningless. One that

---

[1] See any textbook on databases, for example [DATE90].

[2] See Levesque, H., "Knowledge Representation and Reasoning," Annual Review of Computer Science, vol. 1, 1986, for a recent (and orthodox) account of knowledge bases plus a review of the field. [BM86] includes an anthology of conflicting views on knowledge bases, divided along theoretical vs experimental and AI vs Databases party lines.

generates false conclusions (for example, from a / b deduces a) is clearly undesirable. The notions of (logical) soundness and completeness elegantly delimit the class of meaningful and desirable knowledge base operations. Such operations are often called *inferences* because they respect the meaning of their operands.

Knowledge representation system - a generic system for constructing and accessing knowledge bases

Knowledge based system - a computer system which includes in its architecture a knowledge base moreover, the knowledge base is used by other components of the system in a manner that is consistent with its semantics.

Expert system - knowledge based system intended to perform tasks requiring expertise, such as diagnosis, design, planning and interpretation. Expert systems are a special class of knowledge based systems in the sense that not all tasks that might be assigned to a knowledge based system require expertise (consider language understanding, for instance). The focus of expert systems on expertise translates to several requirements on their functionality and operational characteristics, including effective search through large search spaces, an ability to represent and reason about uncertainty, also an ability to explain proposed solutions.[1]

Expert system shell (hereafter ESS) - a generic system offering tools for building and using expert systems. Usually, an ESS includes a knowledge representation system but also explanation, knowledge acquisition, and sophisticated interface facilities.[2]

Knowledge base management system (hereafter KBMS) - a generic system for building and managing a knowledge base. Its features include facilities found in knowledge representation systems for knowledge representation and reasoning but also ones in DBMSs, such as persistence, sharing and efficient access.[3]

Much of the impetus for interfacing knowledge bases and databases, or offering a choice that integrates their functionalities, comes from commercial applications of expert systems. However, the techniques that will be discussed are generally applicable to knowledge based systems.

1.4. The Solution Space

In trying to classify solutions to the problem of interfacing expert systems with databases, it is useful to distinguish between different levels where the interface can be built (Figure 3). The interface may be built for a particular knowledge base - database combination (connecting the top levels of the two towers) or at the levels of the generic systems used to build the two applications that need to be interfaced (leading to a connection of the second levels from the top) or between lower levels, as suggested by the figure.

---

[1] See Hayes-Roth, F., Waterman, D. and Lenat, D. (eds.) *Building Expert Systems*, Addison-Wesley, 1983, for an early but comprehensive collection of papers on expert systems.

[2] For a recent survey of commercial expert system shells see Wang, H., Prager, R. and Fang, J., "Selecting an Expert System Tool: An Empirical Approach," Technical Note CSRI-54, Department of Computer Science, University of Toronto, 1989.

[3] See the volume editted by Mylopoulos and Brodie [BM86] and a companion volume editted by Joachim Schmidt and Costantino Thanos [ST89] for a variety of views on the state-of-the-art in knowledge base management. The volumes summarize the proceedings of two workshops held at Islamorada, Florida and Chania, Crete in 1985, with participation by 30 or so experts on the topics of knowledge bases and databases.

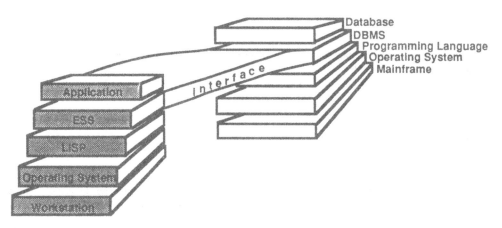

Figure 3

Interfaces between the top levels of the two towers clearly lead to *individualized* or *application-specific* solutions which work for a particular knowledge base - database combination. The XCON/XSEL family of expert systems developed by DEC over the past decade is a prominent example of the individualized solution approach. The development history of XCON shows an ever growing knowledge base and an ever broadening functionality leading to an increased demand for data. When originally deployed in 1979, XCON used 250 rules and handled 100 hardware components. Ten years later, the system used 17,500 rules and handled information on 31,100 components.[1] These demands were met with specific solutions which apply to the XCON hardware/software configuration but generally would not be directly applicable in other contexts.

For obvious reasons, we will focus on generic solutions, shown schematically in Figure 3. After all, a basic premise of this chapter is that direct and efficient access to databases should not be treated as an afterthought by the developers of expert systems. Interfaces at lower levels are also important, but raise few deep technical problems. They will not be discussed here.

Generic solutions can be categorized depending on whether they adopt **existing** generic systems for developing a knowledge base and a database and build an **interface** between them, let's call these *coupling solutions*,[2] and ones which develop an integrated system that supports both expert system functions, such as inferencing, and database management functions, such as persistence and sharing of data. We will refer to this type of solution as an integrated *solution*.

Going a step further, we can classify coupling solutions into ones offering a *loose* or *tight coupling*, depending on the degree of optimization built into the knowledge base - database interface. Likewise, we can further classify integrated solutions into *evolutionary* and *revolutionary* ones, depending on whether they extend an **existing** ESS or DBMS, or start from scratch. Projects working on integrated solutions generally fall under the evolutionary label,

---

[1] It wasn't just the XCON/XSEL knowledge bases and databases that grew. The hardware configuration on which the systems run changed from a single CPU machine to a network of clustered CPUs while the development team grew from 2 to over 50 members. [See Judith Bachant and John McDermott, "R1 Revisited: Four Years in the Trenches," *AI Magazine 5*, (1984) 21-32, for a detailed discussion of the early stages of the project].

[2] See Vassiliou, Y. and Jarke. M. "Coupling Expert Systems with Database Management Systems" [from *Artificial Intelligence Applications for Business*, Ablex Publishing Co, 1984] for one of the first technical treatments of expert system-database interfaces. Among other things, the paper introduces the terms of "loose and tight coupling."

though there have been calls for a revolutionary approach[1] and even some projects that adopt this approach[2] to the problem.

Section 2 of the paper discusses coupling solutions, loose and tight while Section 3 reviews some issues that arise in developing (evolutionary) integrated solutions. Conclusions are presented in Section 4, followed by a selected bibliography. This chapter is non-technical and is intended to give the reader a grasp of the issues instead of technical details of proposed solutions.

2. Coupling Solutions

2.1. Loose Coupling

What are the minimum requirements on an interface between a knowledge base and a database? First, a translation from the data structures used to represent knowledge in the knowledge base -- frames, rules or ground formulas, for example -- to those used by the database -- generally, records or tuples --. Also, we need a translator from expressions accessing the knowledge base to ones accessing the database, and vice versa. Lastly, there has to be a policy on data transfer from one system to the other. Perhaps the simplest policy involves fetching data from the database whenever they are needed.

Figure 4 shows a simplified architecture for a loose coupling between an (expert system) knowledge base and a database.

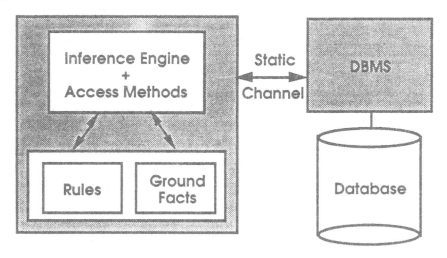

Figure 4

---

[1] See, for instance, opinions expressed in [BM86] and [ST89].

[2] See, for example, Tsur, S. and Zaniolo, C., "LDL: A Logic-Based Data Language," in Proceedings Twelveth International VLDB Conference, Kyoto, August 1986, 33-41.

The coupled system can be visualized here as having two processes, an expert system which is composed of a knowledge base and an inference engine on one hand, and the database system on the other. The two processes exchange messages through a channel of communication which can be said to be "static" in the sense that data transfer is not influenced by the state of the two processes -- for example, the task being carried out by the expert system.

Internally, the expert system knowledge base is assumed to include both generic knowledge ("rules") and ground knowledge (" ground facts"). Data transfer between the DBMS database and the expert system only affects the ground fact component of the knowledge base. Moreover, data are fetched into the knowledge base on an as-needed basis, with no clear policy on when, if ever, revised data are copied back from the expert system database into the permanent store of the DBMS.

Clearly, there are advantages to this type of solution. First, it's simple and easy to conceive and implement. This type of solution is already offered by most ESS vendors coupling their product to a variety of DBMS products.

Unfortunately, loose coupling also has several serious drawbacks.

DBMSs rely heavily on bulk data transfer between main memory and secondary storage to optimize query evaluation with respect to a database. Inference engines, on the other hand, have generally been built with a "find-me-a-solution" mentality[1] -- think how Prolog or a production system searches for a solution --$\sqrt{}$. As an illustration, of the problems that arise, consider so-called recursive queries, involving the recursive use of one or more rules in the knowledge base to achieve a goal. Suppose the expert system needs to find Maria's ancestors given information about people's parents. Loose coupling suggests an evaluation strategy by which the expert system requests and gets first Maria's parents, then the parents of each one of these parents etc., until there are no more ancestors to be fetched from the database. Here the DBMS is being nickeled and dimed to death. Instead, of having the coupled system inherit some query optimization features of DBMSs, it has inherited the naive access methods of the rule interpreter!

And that's not the whole picture! DBMSs pride themselves for supporting effective sharing of data. This means that multiple users may share a single database with a guarantee that these users won't step on each others' toes by affecting each others operations on the database. Moreover, sharing of the database is handled through concurrency control algorithms much more efficient than the obvious solution: locking out all except one user at any one time[2].

Maintaining snapshots of the database elsewhere -- in the ground fact component of the knowledge base -- corrupts this DBMS feature as well. If a system outside the coupling is accessing the DBMS database looking for a particular attribute value, it needs to worry that this value has been updated in the expert system snapshot but hasn't been copied back into the database yet.

Like remarks apply for persistence and error recovery. DBMSs have been endowed with rigorous policies for maintaining a database beyond the execution of applications programs that use it. Also, for restoring a database to a "legal" state in case of a crash. These features aren't standard or even optional for many ESSs. The coupling, again, leads to a least desirable set of features for the composite system.

In summary, loose coupling, though simple to realize, compromises all basic strengths of database technology. For some expert system applications this "naive" access to databases is an acceptable solution.

---

[1] O'Hare and Sheth call this interpreted database access, in contrast to compiled database access performed by DBMSs. See O'Hare, A. and Sheth, A., "The Interpreted - Compiled Range of AI/DB Systems," SIGMOD Record 18, March 1989, for a thorough discussion of compiled vs interpreted database access, including a classification of prototype systems.

[2] A thorough account of concurrency control issues for databases can be found in [BHG87].

That was the bad news. The good news is that depending on the ESS and DBMS you are using, you may be able to go out and buy a coupling solution, as many ESS vendors have strived to bridge the knowledge base - database canyon.[1]

To give you a flavour for the type of solution you may expect to find in the marketplace, consider the KEEconnection, a product offered by IntelliCorp for coupling expert systems developed in KEE, IntelliCorp's ESS, with relational DBMSs[2] (see Figure 5).

A coupling is defined in KEEconnection through a data editor which defines correspondences between relations and their attributes in the database and generic units and their slots in the knowledge base. These are then used to automatically generate database queries from knowledge base expressions.

Data are downloaded from the database transparently when they are needed. Alternatively, the developer of the coupling may define a different downloading strategy when the data set to be downloaded is large and can be characterized in advance. Uploading√ is also possible and is under the full control of the developer. Data communication is handled by software built upon commercial network communications packages.

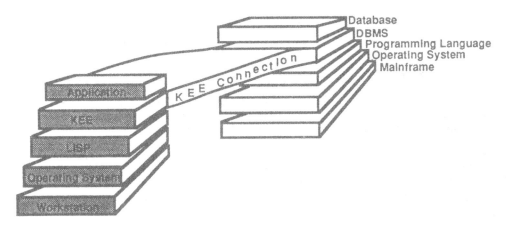

Figure 5

2.2 Tight Coupling

KEEconnection offers a simple downloading/uploading strategy. If this is not suitable for a particular application, the developer is left to his or her own devices (but is at least provided with the programming means to realize them). Tight coupling aspires to offer efficient downloading/uploading techniques, transparent to both the end user and the developer of the coupling.

In a tightly coupled expert system - database architecture, illustrated in Figure 6, the DBMS database can be viewed as an extension of the 'ground facts " component of the knowledge base, managed by a *channel manager* whose job is to optimize data transfer through the channel. The key issue for this architecture is to identify a set of parameters which define this channel optimization. Let us look at two prototype systems which strive to support some form of

---

[1] See report by Ruth Kerry (footnote on page 1) for a table listing several expert system shells and the DBMSs they are coupled with.

[2] Information on the KEE-connection based on IntelliCorp's own publications.

optimization. EDUCE, under development at the European Community Research Centre for more than 5 years [BOCC88] couples a deductive database, consisting of Prolog-like rules, with a conventional relational DBMS.

Expert System

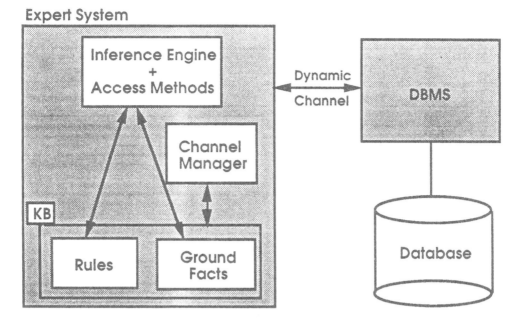

Figure 6

The coupling between the two systems is hybrid in the sense that both loose and tight coupling are supported. The system architecture, shown in Figure 7, has two concurrent processes sharing access to the same database through the same access methods (bottom part of the boxes, representing the two processes in Figure 7).

The loose coupling works as follows: whenever the evaluation of a Prolog goal needs database access, a DBMS query language expression is generated and sent via a pipe to the DBMS. The DBMS evaluates the query and sends back a reply which is further processed to bind Prolog variables to their values. Note that DBMS queries typically return set values which are piped to the Prolog process. Prolog, on the other hand, accepts tuples one at a time, treating the pipe as a queue.

The tight coupling between the two components of EDUCE builds a complete evaluation tree for a Prolog goal and proceeds to evaluate the query; building along the way intermediate relations. Generally, the sets retrieved through this access method from the database are small compared to those retrieved by DBMS queries generated through the loose coupling.

Testing of EDUCE has shown that the tight coupling is faster by one to two orders of magnitude for several different types of queries, particularly recursive ones. There are classes of queries, however, where loose coupling may do better, such as ones involving complex relational algebra expressions.

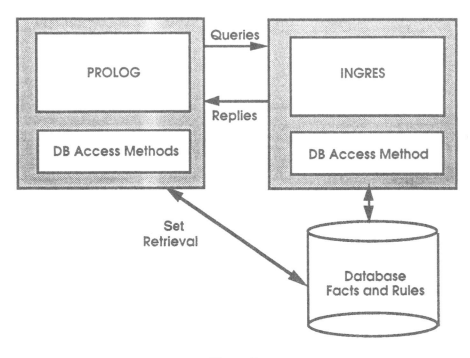

Figure 7

Another lesson learned from EDUCE is that system performance depends critically on the way rules in the knowledge base are represented and stored in the DBMS, in particular whether they are stored in compiled or source form.

EDUCE had to make several changes to Prolog to support its dual coupling to a DBMS. Specifically, loose coupling assumes the user knows both Prolog and the database query language and gives him or her the responsibility of directing commands to the database. BERMUDA [ICFT89] is another experimental prototype system featuring a tight coupling between Prolog and a relational DBMS. Unlike EDUCE, it was designed on the premise that the user wants to see a single interface to the coupled system, Prolog while access of the database is completely transparent.

The system, under development at the University of Wisconsin, assumes many Prolog processes may access a single database through a special module called a BERMUDA agent (Figure 8). The agent receives queries from several Prolog processes and uses several loader processes to fetch data from the DBMS. Queries are sent to the agent when predicates whose extensions are stored in the database need to be accessed. The agent formulates an appropriate query and sends it to an idle loader.The loader fetches the needed data and stores them in a file which is then accessed by the agent when Prolog asks for them.

The BERMUDA agent attempts to optimize data communication with the database through several strategies, including combining several Prolog queries into a single DBMS query, through caching of query answers, also through preemptive data transfer.

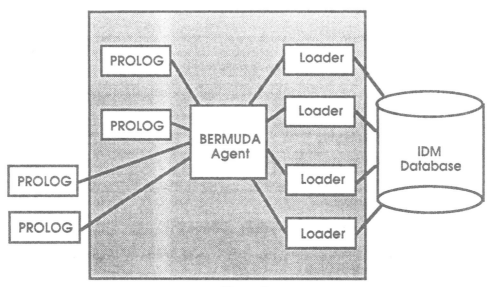

Figure 8

Combining several Prolog queries into a single DBMS query is achieved by looking at the database relations accessed by these queries. Caching of responses to queries, on the other hand, is handled by keeping track of the queries that have been executed and where their answers reside. Prefetching of data from the DBMS is based on the idea that large relations will be fetched into the BERMUDA agent's workspace in pages, which are then used a-tuple-at-a-time by Prolog processes. This means that when a page has been almost used up, the agent can fetch another page of the same relation without waiting for an explicit request from a Prolog process.

Preliminary performance testing shows that, again, tight coupling clearly outperforms a loose Prolog - DBMS coupling, despite the overhead incurred by the BERMUDA agent.

EDUCE and BERMUDA are just two of many experimental systems which attempt to offer a tight coupling between knowledge bases and databases. It should be clear to the reader by now that there is no clear-cut boundary between what we called loosely and tightly coupled systems. Instead, there is a continuum of couplings, determined by the optimization techniques used for data transfer. In short, coupling knowledge bases with databases raises a host of issues, such as:

> **Translation** - the sophistication of the translation process from knowledge base data structures and expressions defined over these to corresponding database data structures and operations.

> **Architecture** - the system structure; the system components (e.g., the rule base, the channel manager, the cache manager) and their interfaces.

> **Semantics** - the treatment of the meaning of the information contained within the two subsystems; interpreting, for instance, missing attribute values from the database.

> **Performance** - the deployment of optimization techniques, the support of recovery mechanisms for subsystem crashes (an interesting feature for, say, real-time expert systems accessing databases)

> **Pragmatics** - many other issues, having to do with the organizational environment within which the coupling takes place.

3. Integrated Solutions

We next turn our attention to integrated solutions to the knowledge base - database interface problem. The idea now is to develop systems which support knowledge representation and reasoning facilities on one hand, and deploy efficient data storage and accessing techniques, like to those offered by DBMSs.

There is no generally accepted method for characterizing knowledge representation and reasoning facilities. For this discussion we shall focus on two types of facilities:

rules and the inference mechanisms needed to use them, such as forward or backward chaining,

structuring mechanisms for a knowledge base, including

generalization - means for building up taxonomies of classes which describe the concepts in the application domain.

aggregation - means for constructing composite descriptions.

classification - means for describing collections of objects (tokens or classes) through the introduction of metaclasses.

We shall use the term "rules" generically to refer to production system-based representations, such as OPS5,[1] and logic-based ones, such as Prolog. After all, similar problems crop up for both types of rules when one aims for integrated solutions.

Turning to database management features for such integrated solutions, the wish list includes most standard DBMS features, such as persistence, sharing, query optimization, concurrency control and recovery.

As shown earlier, integrated solutions can be classified into evolutionary or revolutionary, depending on whether they extend an existing system (ESS or DBMS) or start from scratch. The vast majority of research projects on the topic adopt an evolutionary approach to integration. So, we focus on that approach in the rest of the discussion.

Now, we would have liked to describe a prototype system which attempts to offer all the above features for knowledge representation and reasoning as well as data management. Unfortunately, no such system exists. Instead, we will discuss briefly systems which support either rules or structure and offer (some) database management facilities.

3.1 Rules and Database Management

The integration of rules with database management facilities is a topic of major interest and much research activity today. Some of this interest originated from theoretical research on the relational model which branched out into Logic and Databases and has addressed issues such as recursive query processing[2] [BR86]. A second source of interest originates from DBMS and ESS developers who consider, not without cause, the integration of (production system-like) rules with DBMSs as the shortest path to fame and fortune. We will review a pair of research projects addressing this type of integration and some underlying research issues that need to be addressed.

---

[1] See Forgy, C. and McDermott, J., "OPS: A Domain-Independent Production System Language," Proceedings IJCAI'77, Cambridge MA, 1977.

[2] There have been many surveys of the topic. See, for example, [BR86]

POSTGRES is an experimental extended DBMS which adds OPS5-like rules to an INGRES-like relational DBMS.[1] The system has been under development at the University of California, Berkeley, since the mid '80s. Its features include persistence for rules (not just facts), forward and backward chaining as well as efficient rule selection. Figure 9 shows the architecture of POSTGRES.

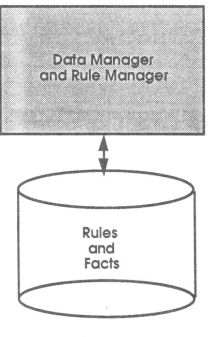

Figure 9

POSTGRES rules are triggered by an event, such as a retrieval or an update operation. For example, the rule

> **ON** replace **TO** EMP.salary **WHERE** EMP.name = "Joe"
> **THEN** replace EMP(salary = NEW.salary)
> **WHERE** EMP.name = "Sam"

will be triggered by updates of the salary attribute of relation EMP where the name attribute of the tuple being updated has value "Joe." Firing this rule causes the salary of the EMPloyee tuple named "Sam" to be updated to the same value as "Joe"'s new salary.

POSTGRES can be seen as a generalized DBMS which, thanks to the presence of rules, supports extended query processing capabilities and a generalized notion of database views. The POSTGRES view mechanism is generalized since it allows, among other things, partial views which are defined both explicitly and through rules.

---

[1] See Michael Stonebraker, Anant Jhingran, Jeffrey Goh, and Spyros Potamianou, "On Rules, Procedures, Caching and Views in Database Systems," Memorandum No. UCB/ERL M89/119, College of Engineering, University of California, Berkeley, October 1989. An earlier description of POSTGRES appears in M. Stonebraker, E. Hanson and C. Hong, "The Design of the POSTGRES Rule System," Proceedings Data Engineering Conference, Los Angeles, 1987.

Efficient query processing relies on two evaluation modes in POSTGRES: direct database access and query modification. Suppose that you have a database with a rule that says that the TOYEMP relation has EMP tuples whose department attribute has value "toy." Further, suppose you want Sam's salary from the TOYEMP relation. The direct database access method will first compute ("materialize") TOYEMP and will find Sam's salary. The query modification method, on the other hand, will change the query, taking into account the definition of TOYEMP, so that the modified query can be evaluated directly against the EMP relation (Figure 10).

Each evaluation strategy is better in some situations and POSTGRES has built-in facilities for determining which strategy to use.

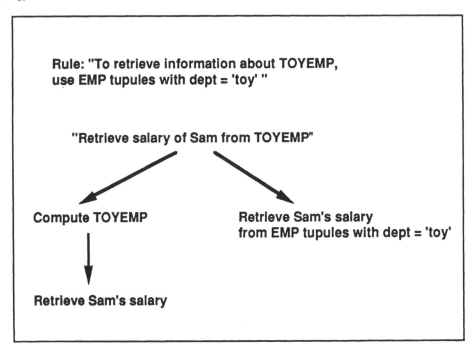

Figure 10

Another interesting prototype is HiPAC,[1] intended to provide a generic tools for applications involving time-constrained database management, such as process control and stock management. The solution offered by HiPAC is based on the notion of an active DBMS, which allows users to specify actions to be taken automatically (without user intervention) when certain conditions arise.

HiPAC offers rules for specifying actions, which, like those of POSTGRES, have an event that triggers the rule, a condition that must be true for the rule to fire and an action. A basic issue addressed in HiPAC is concurrency control for rules. Rules are treated as transactions, atomic database operations which are either successfully executed or quit in which case all their side effects are erased. Since rule firings may cause other rule firings, transactions may be nested, an old and thorny issue in database management.

---

[1] See Dennis McCarthy and Umeshwar Dayal, "The Architecture of an Active Database Management System," Proceedings SIGMOD Conference, Portland, May 1989, 215-224.

There are important differences in emphasis between HiPAC and POSTGRES rules. Unlike POSTGRES which has focused on efficient firing and evaluation of possibly recursive rules, HiPAC concentrates on time-constrained scheduling of database transactions.

HiPAC is based on the PROBE data model [MD86] which is functional/object-oriented and offers facilities for modelling temporal and spatial data. Its implementation is being carried out in Smalltalk.

Figure 11 shows the architecture of a HiPAC application. The system shown is intended to aid a securities analyst by presenting her with information of interest and by executing trades according to the analyst's instructions.

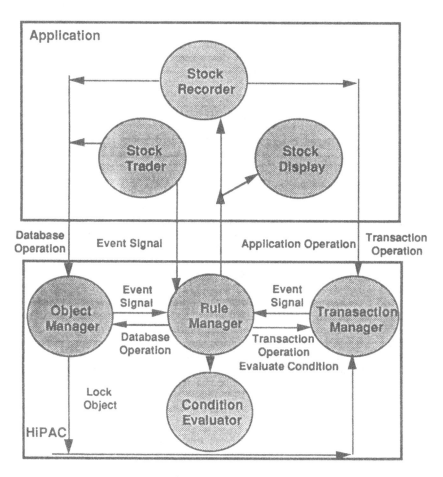

Figure 11

The three modules constituting the application update current prices in the system database (Stock recorder module), display prices, trades, portfolios etc (Stock display module) and execute trades (Stock trade module).

The components of HiPAC include an object manager, which provides object-oriented database management, a transaction manager, which supports nested transactions, a rule manager which maps events to rule firings and rule firing to transactions, and lastly a condition evaluator which evaluates rule conditions to determine whether a rule is eligible for firing.

What are the dominant issues that have arisen from research projects such as POSTGRES and HiPAC? As the reader might have expected, the efficient implementation of rules heads the list of concerns. The obvious implementation method for rule selection, iterating for each rule through all facts to find out whether the rule can fire, is simply too costly for large rule and fact bases. **Indexing** of rules and facts in the database with predicates reduces matches that will be tried for each rule and is therefore, a step in the right direction. **Incremental access methods** are a second class of optimizations which attempt to use caching of previous computations to avoid costly recomputations. Incremental access methods have been proposed in at least two different forms. Rete-type algorithms[1] maintain information about the status of a match between each rule condition and the fact database on a network structure. This information is used to determine whether a database update affects a rule match or not without having to recompute all rule matches. The other incremental access method originates from database research. The basic idea is to cache the result of a query evaluation -- possibly a very large data set -- for later use rather than discard it.[2] The effectiveness of this view materialization is enhanced by supplying algorithms which given a database update, determine how a materialized view should be changed to remain consistent with the updated version of the database.

A second important concern in implementing rules is concurrency control. A query and an update are logically independent when the latter does not affect the value of the former, in which case they can be executed concurrently. Unlike conventional databases where logical independence can be determined algorithmically by inspecting the query and update in question, it can be shown that the problem of logical independence between a knowledge base query and an update is undecidable, even for knowledge bases having ground and Horn clauses only. However, some results are beginning to appear on interesting special cases of the problem which admit tractable solutions.[3]

As we showed above, integrating rules with database management facilities is largely a research topic. However, some DBMS vendors (e.g., Ingres) are beginning to offer some form of integrated rule processing for their products. While Deductive DBMSs are becoming commercially available, it is unlikely that they will be able to compete against relational DBMSs extended with rule processing capabilities.

3.2 Structuring Mechanisms and Database Management

We next turn to the integration of knowledge structuring facilities, such as aggregation and generalization, with database management facilities. Consider aggregation first. It is well-known that relational databases fare badly when they need to represent complex objects, such as ones encountered in CAD applications, because they insist that complex descriptions be broken down into elementary components. Moreover, this failure is both in the expressiveness of the relational data model -- it's just plain hard to represent complex objects with tuples -- and in performance -- it's just plain costly to disassemble and reassemble complex descriptions.

---

[1] See Forgy, C. L. "Rete: A Fast Algorith for the Many Pattern/Many Object Pattern Match Problem," *Artificial Intelligence 19*, 1982, 17-37.

[2] See, for instance, Nicholas Roussopoulos, "THe Incremental Access Method of View Cache: Concepts, Algorithms and Cost Analysis," CS-TR-2193, Department of Computer Science, University of Maryland, 1989.

[3] For example, see Charles Elkan, "Independence of Logic Database Queries and Updates," Proceedings PODS-90.

Solutions to the problem of supporting aggregation come in different forms. Some extensions of relational DBMSs allow nesting of tuples.[1] At a different level, object-oriented representations do away altogether with the notion of tuple and replace it with that of object.[2] We'll discuss solutions of the latter type in two stages.

First, consider the facilities provided for describing aggregate objects. To represent the components of an object, one can use derived representations through expressions evaluated whenever the components need to be fetched; alternatively, one can use identifiers which name explicitly the components of an object; or one can insist on fully expanded descriptions of an object, *cached representations* if you like.

Turning to performance issues, clearly there is a time-space tradeoff between derived and cached representations, with identifier-based representations falling somewhere between. Implementation of identifiers can be achieved through physical caching, but then it is as space-expensive as cached representations, or through physical clustering techniques which sometimes can be more efficient with respect to both time and space.[3]

Generalization comes next. As a knowledge structuring facility, it has offered taxonomies as means for coping with large numbers of like concepts. From an implementation point of view, the list of issues that need to be addressed includes the physical treatment of inheritance -- do you store or compute inherited attributes -- and the physical layout of the sets of instances of each class (or concept or frame).[4] The two basic approaches to the later problem involve so-called vertical and horizontal partitioning.

To illustrate the relative advantages of the two approaches, consider a simple knowledge base of *Persons, Students* and graduate students or *Grads* (Figure 12). In a vertical partitioning this knowledge base, attributes are grouped and each group stored separately. For example, attributes can be stored with instances of the class in which they defined. The *Persons* attributes of a graduate student are stored in Persons instances, separately from *Students* attributes and the Grads attributes. Horizontal partitioning, on the other hand, stores all attributes of an object together but partitions the instances of the class *Persons* to those which are also students and those which are in addition graduate students. Clearly, iteration over all instances of a class can be done more efficiently with vertical instead of horizontal partitioning. Fetching all attributes of an object, on the other hand, can be done more efficiently with horizontal partitioning.

---

[1] Hans-Joerg Schek, Heinz Bernhard Paul, Marc Scholl and Gerhard Weikum, "The DASDBS Project: Objectives, Experiences and Future Prospects," *IEEE Transactions on Knowledge and Data Engineering 2*, March 1990, describes a research project which focuses on the efficient implementation of nested relations.

[2] Several other types of database systems offer extensions to classical DBMSs which accomodate some form of structure consistent with the requirements of knowledge representation:

semantic data models: extensions of classical data model which support the notions of entity and relationship, as well as generalization, aggregation and classification.

special data models: these are extensions which offer facilities for representing temporal or spatial data

data base programming languages: like object-oriented databases, they strive to offer database management and programming language facilities within one framework

[3] For more details on these alternatives see Anant Jhingran and Michael Stonebraker, "Alternatives in Complex Object Representation: A Performance Perspective," Memorandum No. UCB/ERL M89/18, College of Engineering, University of California, Berkeley, February 1989.

[4] A good case study of the tradeoff issues that arise in implementing efficiently generalization hierarchies appears in [NCLB87]. See also Grant Weddell, *Physical Design and Query Compilation for a Semantic Data Model*, PhD thesis, Department of Computer Science, University of Toronto, 1987, for elegant results on the complexity of finding an optimal physical layout for genralization hierarchies.

VERTICAL SPLITTING

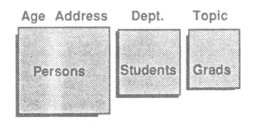

HORIZONTAL SPLITTING

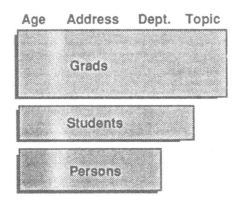

Figure 12

We will not discuss particular commercial or experimental systems which support knowledge structuring and database management facilities. There over ten commercial object-oriented database systems[1] -- and their number keeps changing by the day -- which do support aggregation and generalization in some form. Generally, however, it seems that the object-oriented database systems are focusing on database programming applications and should be seen as an interesting parallel technology to knowledge base management instead of part of the solution.

One interesting result of the research effort to develop object-oriented database systems is a consensus on the basic architecture that is needed. This architecture has two components, a storage manager supporting a simple data model and responsible for resource allocation and concurrency control, also an execution layer which supports a more sophisticated data model (programming) language facilities and does workspace management. This architecture is similar but less ad hoc to ones that have been adopted for the implementation of ESSs offering knowledge structuring and inferencing facilities.

Commercial object-oriented database systems notwithstanding, there are several research issues that need to be addressed in integrating knowledge structure and management. Concurrency control issues are a case in point.

---

[1] The September 1989 issue of a newsletter titled Release 1.0, presents a recent and thorough survey of object-oriented database system products, available or in prototype form. For expert system applications, Statice sold by Symbolics is of particular interest, not only because of its vendor. Statice was one of the first Lisp-based object-oriented database system to be commercially available. Gemstone, the most widely used, is Smalltalk-based.

Aggregation complicates the atomic units out of which a knowledge base is built. Generalization implies that class extensions may not be mutually exclusive. Since a knowledge base operation, say addition of some knowledge to the knowledge base, involves enforcing inherent generalization and aggregation constraints, it can hardly be treated as atomic. These facts suggest that the results of concurrency control theory need to be extended before they can be applied to knowledge base management.

Little is known, theoretically or experimentally about the factors that affect performance for systems which support structure and management facilities (let alone **rules**, structure, and management facilities).

Lastly, it seems clear that a system supporting both knowledge structuring and management facilities needs to include in its facilities version control for knowledge base objects (as well as knowledge bases), plus facilities for configuration management -- which versions of objects are compatible in forming larger assemblies -- and schema evolution -- what happens to the instances of a generic description that has just been changed.

4. Summary and Future Directions

In summary then, here is the state-of-the-art in interfacing knowledge bases with databases. Loose coupling is commercially available and will be sufficient for many applications calling for expert system - database interaction. Tight coupling is still under study in research laboratories around the world, but should be commercially available within 2 - 3 years. Integration of rules or knowledge structuring facilities with management facilities is still being addressed at the research prototype stage but again will soon be commercially available. For example, Ingres currently offers a Rule Manager. Lastly, deep integration of a full set of knowledge representation and reasoning facilities on one hand and management facilities on the other should not be expected for some time to come, perhaps for as long as the end of the decade.

Integrating knowledge bases and databases can be very hard indeed, depending on the required sophistication of a solution. But, this is not the only message for the reader. A key research issue for knowledge representation and reasoning in the last few years has been the study of expressiveness - tractability tradeoffs for (knowledge representation) formalisms. This research perspective is based on theoretical (usually worst case) analyses of algorithmic complexity of knowledge base operations. Complexity theory has finally met knowledge representation, implying an admission on our part that there is no point in designing knowledge representation schemes that can be shown to be computationally intractable. To build systems which support knowledge representation, reasoning **and management** facilities, this research will have to be extended to a theory of how do knowledge based systems perform on particular machine architectures and with specific assumptions of knowledge base size and usage. We'll have to develop a performance theory of knowledge based systems if we shall succeed in developing a technology for building systems that can interpret and manage large amounts of information.

It is impossible to discuss trends on the topic of knowledge bases and databases without having a vision of the future and the information systems it will demand. Let us consider an example involving health workers in a hospital. There are many such workers, e.g., doctors, nurses, technicians and administrators, each with their own knowledge and reasoning facilities, possibly augmented by computerized information systems, cooperating to heal a patient. Consider a patient, for example, who visits his physician, then a radiology lab and a cardiologist, and finally the billing department of a hospital (Figure 13). Obviously, there has to be much communication and cooperation among the different agents to help the patient and keep the complete system functioning. The knowledge each has often involves specialized areas, such as cardiology or radiology. The knowledge used by different agents may be overlapping, consistent or inconsistent and you may want to maintain the inconsistencies. How do these agents interact and cooperate? How can the **computer systems** they use be made to communicate and cooperative, even if they were built independently? [BBHL88]

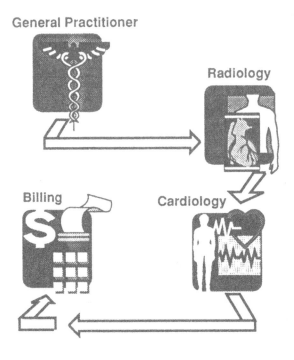

Figure 13

We believe that the paradigm for the next generation of information systems (circa 1995-2000) will involve large numbers of intelligent agents distributed over large computer/communication networks. Agents will include humans, humans working with computer systems, and computer systems performing tasks without human intervention. Work will be conducted on the network in many forms. Work tasks will be defined by one (centralized -- e.g., a complex engineering task) or more (decentralized -- e.g., monitoring many systems of a patient or many stations in a factory) agents and will be executed by agents acting autonomously, cooperatively, or collaberatively, depending on the resources needed to complete the task. Agents will request, and get, resources (e.g., processing, knowledge, data) without knowing what resources are required, how those resources will be acquired, and how the resources will be orchestrated to achieve the desired result. A goal of this vision of the future is to be able to efficiently, and transparently, use all computing resources available on all computers in large computer/communications network.

The design, construction, use, and evolution of systems within the above paradigm will require sophisticated support for all aspects of the systems life cycle. We call such systems **Intelligent Information System** (IISs).

The concept of an IIS is evolving from several currently disjoint technologies, including those that have given us knowledge bases and databases. It is easy to conceive of powerful knowledge based systems (e.g., medical diagnostic systems) with efficient access to several large information bases (e.g., databases, files). It is also easy to imagine large information systems (e.g., airline reservation systems) with added "intelligence" (e.g., scheduling and re-scheduling trips based on changing conditions or considering individual's preferences). IISs will be considerably more powerful than such simple extrapolations of existing systems concepts. Because of a lack of familiarity with the new computing paradigm, it is not easy to imagine potential IISs that would take advantage of the features proposed above. Academic and industrial researchers are just beginning to play with these notions in areas such as Computer Integrated Manufacturing, Office Automation, Collaborative Work, Coordination Technology, Groupware (Intelligent) Interoperability, Intelligent (Communication) Networks, Distributed AI, Human-Computer Interaction and many others.

Unlike past major advances in computing, the next generation of information systems (whatever its form) depends on the integration of currently disjoint technologies. **Database Systems** can contribute with information management techniques, particularly for distributed or heterogeneous databases, and efficient implementation techniques for information bases. **Artificial Intelligence** can contribute with knowledge representation and reasoning technique, on one hand, and distributed problem solving in a multiagents environment on the other. **Operating Systems** can contribute with resource management techniques over a large distributed computer/communications network. **Programming Languages** can contribute with languages and type/object systems for concurrent programming. **Software, Knowledge, and Data Engineering** can each contribute with design and development environments, shells and methodologies for building such IISs. **Computer Communications** can provide the necessary underlying communication and interconnection technology. These are only a few of the component technologies and the contributions they can potentially make to the topic at hand.

5. Selected Bibliography

[ABRI74]
> Abrial, J.R., "Data Semantics," in *Data Management Systems*, J.W. Klimbie and K.L. Koffeman (eds.), North Holland, 1974.

[ACO85]
> Albano, A., L. Cardelli and R. Orsini, "Galileo: A Strongly Typed, Interactive Conceptual Language," *ACM Transactions on Database Systems 10*, No. 2, March 1985.

[ASTR76]
> Astrahan, M. M., et. al., "System R: Relational Approach to Database Management," *ACM Transactions on Database Systems*, 1, 1976.

[BANC86]
> Bancilhon, F., "Naive Evaluation of Recursively Defined Relations," in [BM86].

[BATO86[
> Batory, D.S., et. al., "GENESIS: A Reconfigurable Database Management System," Technical Report TR-86-07, University of Texas, Austin, 1986.

[BBBD88]
> Bancilhon, F., G. Barbedette, V. Benzaken, C. Delobel, S. Gamerman, C. Lecluse, P. Pfeffer, P. Richard and F. Velez, "The Design and Implementation of O2, and Object-Oriented Database System," in [OODB88].

[BBGS86]
> Batory, D.S., J.R. Barnett, J.F. Garza, K.P. Smith, K. Tsukuda, B.C. Twichell and T.E. Wise, "GENESIS: An Extensible Database Management System," *IEEE Transactions on Software Engineering* (to appear); also technical report 86-07, Department of Computer Science, University of Texas at Austin, 1986.

[BBHL88]
> Brodie, M.L., D. Bobrow, C. Hewitt, V. Lesser, S. Madnick and D.C. Tsichritzis, "Future Artificial Intelligence Requirements For Database Systems," in [KERS88].

[BEER88]
> Beeri, C. (ed.), *Proc. Third International Conference on Data and Knowledge Bases,* Israel, 1988.

[BGM85a]
> Bouzeghoub, M., G. Gardarin, E. Metais, "Database Design Tools: An Expert System Approach," *Proc. International Conference on Very Large Databases,* Stockholm, 1985.

[BHG87]
> Bernstein, P.A., V. Hatzilakos and N. Goodman, *Concurrency Control and Recovery in Database Systems,* Addison-Wesley, Reading, MA, 1987.

[BL86]
    Brachman, R.J., H.J. Levesque, "What Makes a Knowledge Base Knowledgeable? A View of Databases from the Knowledge Level," in [KERS86], 69-78.

[BL86a]
    Brachman, R. and H. Levesque, "The Knowledge Level of KBMSs," in [BM86].

[BZ87]
    Bloom, T. and S.B. Zdonik, "Issues in the Design of Object-Oriented Database Programming Languages," in [MEYR87].

[BM86]
    Brodie, M. and J. Mylopoulos, *On Knowledge Base Management Systems: Integrating Artificial Intelligence and Database Technologies*, Springer-Verlag, New York, 1986. (*Note:* This volume includes the proceedings of the Islamorada workshop held in February 1985 in Islamorada, FL.)

[BM86a]
    Brodie, M.L. and J. Mylopoulos, "Knowledge Bases and Databases: Semantic vs. Computational Theories of Information," in [AC86].

[BMS84]
    Brodie, M.L., J. Mylopoulos and J.W. Schmidt (eds.), *On Conceptual Modelling: Perspectives from Artificial Intelligence, Databases, and Programming Languages*, Springer-Verlag, New York, 1984

[BOCC88]
    Bocca, J., "On the Evaluation Strategy for EDUCE," Proceedings, SIGMOD-86, Washington, May 1986,368-378. Reprinted in Mylopoulos, J. and Brodie, M. (eds.) Readings in Artificial Intelligence and Databases, Morgan Kaufmann, 1988.

[BR84]
    Brodie, M.L. and D. Ridjanovic, "On the Design and Specification of Database Transactions," in [BMS84], revised and republished in [MB88].

[BR86]
    Bancilhon, F. and R. Ramakrishnan, "An Amateur's Introduction to Recursive Query Processing," *Proc. ACM SIGMOD International Conference on Management of Data*, Washington, D.C., May 1986.

[BROD84]
    Brodie, M.L., "On the Development of Data Models," in [BMS84].

[BROD86]
    Brodie, M.L., "Knowledge Base Management Systems: Discussions from the Working Group," in [KERS86].

[BZ81]
    Brodie, M.L. and S.N. Zilles (eds.), *Proc. Workshop on Data Abstraction, Databases and Conceptual Modelling*, Joint Special Issue of *SIGPLAN Notices, SIGMOD Record*, and *SIGART Newsletter*, January 1981.

[CARE86]
    Carey, M.J., et. al., "Object and File Management in the EXODUS Extensible Database System," *Proc. Twelfth International Conference on Very Large Data Bases*, August 1986.

[CDFG86]
    Carey, M.J., D.J. DeWitt, D. Frank, G. Graefe, M. Muralikrishna, J.E. Richardson and E.J. Shekita, "The Architecture of the EXODUS Extensible DBMS," in [DD86], 52-65.

[CHEN76]
    Chen, P.P. -S., "The Entity-Relationship Model: Towards a Unified View of Data," *ACM Transactions on Database Systems 1*, No. 1, March 1976.

[CODD70]

Codd, E.F., "A Relational Model for Large Shared Data Banks," *Communications of the ACM 13*, No. 6, June 1970, 377-387.

[CODD79]

Codd, E.F., "Extending the Database Relational Model to Capture More Meaning," *ACM Transactions on Database Systems 4*, No. 4, December 1979.

[DATE90]

Date, C.J., *An Introduction to Database Systems*, 5th edition, Addison-Wesley, Reading, MA, 1990.

[DATE83]

Date, C.J., *An Introduction to Database Systems*, Volume II, Addison-Wesley, Reading, MA, 1983.

[DD86]

Dittrich, K.R., and U. Dayal (eds.), *Proc. 1986 International Workshop on Object-Oriented Database Systems*, Washington, IEEE Computer Society Press, 1986.

[DGL86]

Dittrich, K.R., W. Gotthard and P.C. Lockemann, "DAMOKLES--A Database System for Software Engineering Environments," *Proc. IFIP Workshop on Advanced Programming Environments*, R. Conradi, T.M. Didriksen and D.H. Wanvik (eds.), Trondheim, Norway, June 1986, Lecture Notes in Computer Science 244, Springer-Verlag.

[DS86]

Dayal, U. and J.M. Smith, "PROBE: A Knowledge-Oriented Database Management System," in [BM86].

[GALL83]

Gallaire, H., "Logic Databases vs. Deductive Databases," *Logic Programming Workshop*, Albufeira, Portugal, 1983, 608-622.

[GD87]

Graefe G. and D.J. DeWitt, "The EXODUS Optimizer Generator," *Proc. ACM SIGMOD International Conference on Management of Data*, May 1987.

[GLP75]

Gray, J.N., R.A. Lorie, and G.R. Putzolu, "Granularity of Locks in a Shared Database," *Proc. First International Conference on Very Large Databases*, Framingham, MA, 1975, 428-451.

[GM78]

Gallaire, H. and J. Minker (eds.), *Logic and Databases*, Plenum Press, New York, 1978.

[GM88]

Graefe, G. and D. Maier, "Query Optimization in Object-Oriented Database Systems: A Prospectus." In [OODB88].

[GMN84]

Gallaire, H., J. Minker and J. Nicolas (eds.), *Advances in Database Theory*, Plenum Press, New York, 1984.

[GMN84a]

Gallaire, H., J. Minker and J. Nicolas (eds.), "Logic and Databases: A Deductive Approach," *ACM Computing Surveys 16*, No. 2, June 1987.

[HK87]

Hull, R. and R. King, "Semantic Database Modelling: Survey, Applications and Research Issues," *ACM Computing Reviews 19*, No. 3, September 1987.

[HM81]

Hammer, M.M. and D.J. McLeod, "Database Description with SDM: A Semantic Database Model," *ACM Transactions on Database Systems 6*, No. 3, September 1981.

[HM85]

Heimbigner, D. and D. McLeod, "A Federated Architecture for Information Management," *ACM Transaction on Office Information Systems 3*, No. 3, July 1985, 253-276.

[ICFT89]

Ioannidis Yannis, Joanna Chen, Mark Friedman, Manolis Tsangaris, "BERMUDA -- An Architectural Perspective on Interfacing Prolog to a Database Machine" in Kerschberg, L. (ed.) *Expert Database Systems*, Proceedings of the 1988 International Conference, The Benjamin Cummings Publishing Co., 1989.

[KB85]

Ketabchi, M.A. and V. Berzins, "ODM: An Object-Oriented Data Model for Design Databases," Univ. of Minnesota Institute of Technology TR 85-41, October 1985.

[KENT78]

Kent, W., *Data and Reality*, North-Holland, Amsterdam, 1978.

[KENT79]

Kent, W., "Limitations of Record-Based Information Models," *ACM Transactions on Database Systems 4*, No. 1, 1979, 107-131.

[KERS86]

Kerschberg, L. (ed.), *Expert Database Systems: Proc. from the First International Workshop*, Benjamin/Cummings, Menlo Park, CA, February 1986.

[KERS87]

Kerschberg, L. (ed.), *Expert Database Systems: Proc. from the First International Conference*, Benjamin/Cummings, Menlo Park, CA, 1987.

[KERS88]

Kerschberg, L. (ed.), *Expert Database Systems: Proceedings from the Second International Conference*, Benjamin/Cummings, Menlo Park, CA, 1988.

[KING83]

King, J.J. (ed.), *Special Issue on AI and Database Research, ACM SIGART Newsletter*, October 1983.

[KL88]

Kim W. and F. Lochovsky (eds.), *Object-Oriented Languages, Applications, and Databases*, Addison-Wesley, Reading, MA (to appear).

[KM85]

King, R. and D. McLeod, "Semantic Data Models," in *Principles of Database Design, Volume I: Logical Organizations*, S.B. Yao (ed.), Prentice-Hall, New York, 1985.

[KRB85]

Kim, W., D. Reiner and D. Batory (eds.), *Query Processing in Database Systems*, Springer-Verlag, New York, February 1985.

[LMP87]

Lindsay, B., J. McPherson and H. Pirahesh, "A Data Management Extension Architecture," *Proc. ACM SIGMOD International Conference on Management of Data*, May 1987, 220-226.

[LR82]

Landers, T. and R. Rosenberg, "An Overview of MULTIBASE," *Proc. Second International Symposium on Distributed Databases*. Berlin, West Germany, September 1982.

[MB88]

Mylopoulos, J. and M.L. Brodie, *Readings in Artificial Intelligence and Databases*, Morgan Kaufmann, San Mateo, CA, 1988.

[MBW80]

Mylopoulos, J., P. Bernstein and H.K.T. Wong, "A Language Facility for Designing Database-Intensive Applications," *Transactions on Database Systems 5*, No. 2, June 1980, 185-207.

[MD86]

Manola, F. and U. Dayal, "PDM: An Object-Oriented Data Model," in [DD86].

[MW80]

Mylopoulos, J. and H. Wong, "Some Features of the TAXIS Data Model," *Proc. Sixth International Conference on Very Large Databases*, Montreal, October 1980.

[NCLB87]

Nixon, B., L. Chung, D. Lauzon, A. Borgida, J. Mylopoulos and M. Stanley, "Implementation of a Compiler for a Semantic Data Model: Experiences with Taxis," *Proc. ACM SIGMOD International Conference on Management of Data*, 1987.

[OODB88]

Advanced in Object-Oriented Database Systems: Proc. Second International Workshop on Object-Oriented Database Systems, K. Dittrich (ed.), Bad Muenster am Stein-Ebernburg, West Germany, September 1988, Lecture Notes in Computer Science 334, Springer-Verlag.

[PM88]

Peckham, J. and F. Maryanski, "Semantic Data Models," *ACM Comuting Surveys*, Vol. 20, No. 3, September 1988, 153-190.

[REDD88]

Reddy, R., 'Foundations and Grand Challenges of Artificial Intelligence ", AAAI Presidential Address '88, *AI Magazine*, Winter 1988, 9-21.

[ROUS76]

Roussopoulos, N., *A Semantic Network Model of Databases*, Ph.D. dissertation, Department of Computer Science, University of Toronto, Toronto, 1976.

[RS87]

Rowe, L. and M. Stonebraker, "The POSTGRES Data Model," *VLDB XIII*, Brighton, England, September 87.

[SFL83]

Smith, J.M., S.A. Fox and T. Landers, "ADAPLEX Rationale and Reference Manual," Technical Report CCA-83-08, Computer Corporation of America, Cambridge, May 1983.

[SHIP81]

Shipman, D., "The Functional Data Model and the Data Language DAPLEX," *ACM Transaction on Database Systems 6*, No. 1, March 1981, 140-173.

[SIBL76]

Sibley, E.H. (ed.), "Special Issue: Data-Base Management Systems," *ACM Computing Surveys, 8*, No. 1, March 1976.

[SMIT81]

Smith, J.M., et. al., "MULTIBASE -- Integrating Heterogeneous Distributed Databases," *Proc. AFIPS National Computer Conference, 50*, June 1981.

[SR86]

Stonebraker, M. and L.A. Rowe, "The Design of POSTGRES," *Proc. ACM SIGMOD International Conference on Management of Data*, Washington, D.C., May 1986, 340-355.

[SS77a]

Smith, J.M. and D. Smith, "Data Abstraction: Aggregation and Generalization," *ACM Transactions on Database Systems 2*, No. 2, June 1977, 105-133.

[ST89]

Schmidt, J.W. and C. Thanos (eds.), *Fundamentals of Knowledge Base Management Systems*, Springer-Verlag, New York, 1988.

[STON75]

Stonebraker, M., "Implementation of Integrity Constraints and Views by Query Modification," *Proc. ACM SIGMOD International Conference on Management of Data*, San Jose, CA, May 1975.

[STON86b]

Stonebraker, M. and M.R. Stonebraker, "Inclusion of New Types in Relational Database Systems," *Proc. Second International Conference on Data Base Engineering,* Los Angeles, February 1986.

[ULLM88]

Ullman, J.D., *Principles of Database and Knowledge-Base Systems,* Volume I, Computer Science Press, Potomac, MD, 1988.

[ZANI83]

Zaniolo, C., "The Database Language GEM," *Proc. ACM SIGMOD Conference on Management of Data,* San Jose, CA, May 1983.

[ZM89]

Zdonick, S.B. and D. Maier, *Readings in Object-Oriented Databases,* Morgan Kaufmann, San Mateo, CA, 1989.

Terminological Reasoning and Information Management*

Bernhard Nebel

Deutsches Forschungszentrum
für Künstliche Intelligenz
D-6600 Saarbrücken

Christof Peltason

Technische Universität Berlin
Project KIT-BACK
D-1000 Berlin 10

Abstract

Reasoning with terminological logics is a subfield in the area of knowledge representation that evolved from the representation language KL-ONE. Its main purpose is to automatically determine the location of a new concept description (or object description) in a partially ordered set of given concepts. It seems to be a promising approach to apply the techniques developed in this area to the development of new object-based database models. The main advantages are a uniform query and database definition language and the utilization of an indexing technique, which we call semantic indexing.

1 Introduction

The development of elaborate techniques for information description is an important task in building advanced information systems. The appropriate means of describing classes, objects, and complex dependencies of an application domain can help users to express their problems in a natural way and make an important contribution to the effort of turning an information management system into a system which finally might be called a system managing a "knowledge base". In addition, description techniques can also be exploited to guide the internal reasoning and retrieval processes of the information management system. The main goal of this paper is to

---

*This work was supported by the German Ministry for Research and Technology BMFT under contract ITW 8901 8, and by the Commission of the European Communities within ESPRIT Project 311.

show that a proper treatment of the domain terminology is a good starting point for dealing with both aspects, the usage aspect of the system's expressiveness and the implementation aspect of information management.

We will start in Section 2 with a short overview of the underlying *terminological logic* approach. This representation paradigm has evolved from the work in the context of the knowledge representation language KL-ONE [12], and has gained a wide audience during the last decade. We will show how to use a terminological description language for the modelling of domain entities, and how to build expressions for complex information retrieval tasks within this framework. This exposition will use the formalism employed in the BACK system [42], thereby introducing an essential subset of the BACK-formalism.

In Section 3 we will sketch some aspects of implementing an information system following this approach. We discuss how management and persistency procedures can take advantage of a knowledge base which is structured by a terminological scheme.

The interdependencies between the related work in knowledge representation and advanced database systems are sketched in Section 4. One interesting result of this survey is that database research can profit from research in knowledge representation. In particular, we will point out that the *refinement* algorithm for the object-oriented database system O_2 published in [27] is incomplete and argue that the problem itself is intractable – insights based on theoretical results achieved in the area of terminological logics.

While the approach we focus on in this paper stems from a tradition in Artificial Intelligence research it is interesting to see how it has gradually also become part of the converging tendencies between research on databases and AI (cf. [13]). Originally, the notion of a knowledge representation system was often used in its most ambitious variant, i.e. as an attempt to support the representation of all aspects of knowledge, such as dynamic processes, various kinds of natural-language phenomena, uncertain or vague information, beliefs, and many more. The paradigm of terminological reasoning, however, started from a very limited (but well-founded, set-theoretical) formalism. Although the limitation led to an increasing distance from the initial, purely AI-oriented goals, the results were acknowledged as contributions to the research area of database systems, where they now are attracting a growing interest. The reason is that – due to the formal rigidity of the approach – the behaviour of the systems can be estimated in a reliable way and the representational service they provide can be seen as playing a central role within any advanced information management system.

In order to sketch a first, intuitive picture of the idea we look at the following scenario:

Let us assume we are talking about organizations, i.e. universities, companies, and research institutes applying for research projects, or more precisely *Esprit* projects. The legal status of these organizations is determined by their locations (or rather where they have their formal residences), by the number of employees, etc., which finally makes them eligible for Esprit projects and constitutes Esprit Consortia. What kind of interaction features would we expect from an "intelligent" information system?

First, a support for building a formal model of this application domain, i.e. for describing the abstract entities such as *universities* or *CEC-companies* is required. We may also need a control of the dependencies between these descriptions in the modelling phase in order to be able to estimate the reasoning processes in the next phases. Then we would like to enter information about concrete objects in our domain such as:

> *The German Research Center for AI and the Technical University Berlin are members of the Esprit Project 42 Consortium.*

No deeper, implementation dependent knowledge of the underlying scheme should be required for such entries. All information should be describable in a logic-oriented way. Finally, we would like to retrieve our information in as comfortable a manner as possible:

> *What are the Esprit Consortia which have only small and medium enterprises (SMEs) as their members?*

On the whole, entries and queries should be describable in a logic-oriented way using complex descriptions of *what* we want to know rather than requiring references as to *where* to find it. In the following sections we will see how this is reflected in a uniform approach for knowledge base access.

2 The Terminological Reasoning Approach

The terminological approach is based on a clear distinction between intensional and extensional descriptions. Although originally introduced as an epistemological category, the distinction turns out to be useful from a technical information processing point of view as well. It offers a clean methodology for distinguishing between the level for reasoning about abstract classes and the level for reasoning about objects which instantiate these classes. On the one hand, this brings the notion close to the conventional database-like distinction between *database scheme* and *database extensions*. On the other hand, however, the language designed for intensional descriptions (which could be called a *knowledge base scheme language*) is an expressive language within which complex descriptions can be built, thus constituting a highly complex data model which – for a semantically well-founded language – includes a complex reasoning machinery.

2.1 Term Description Language

The language designed for intensional descriptions is called the *Term Description Language*. It contains a repertoire of constructs which may vary among the different incarnations of this system family, however, in its core at least the following can be identified:

Terminology

Intensional descriptions are introduced as *term equations* with *names* as left-hand sides and composite *terms* as right-hand sides. There are two kinds of introductions, *primitive* ones, indicated by :<, and *defined* ones, indicated by :=, an alternative we will explain further below. A sequence of such introductions makes up the *terminology*, or the *terminological model* of a domain.[1]

$$
\begin{aligned}
\langle term\text{-}tell\rangle \quad ::= \quad & \langle concept\text{-}\textsc{name}\rangle :< \langle concept\rangle \\
| \quad & \langle concept\text{-}\textsc{name}\rangle := \langle concept\rangle \\
| \quad & \langle role\text{-}\textsc{name}\rangle :< \langle role\rangle \\
| \quad & \langle attribute\text{-}set\text{-}\textsc{name}\rangle := \langle attribute\text{-}set\rangle
\end{aligned}
$$

For building terms we distinguish between *classes* and *roles* where *classes* denote sets and *roles* denote relations between these sets. Among the different kinds of classes we use *concepts* for intensional descriptions which – by virtue of semantically based subsumption – form an abstraction hierarchy:

---

[1]The semantics of the term description language is given in the Appendix.

$$\langle term \rangle \quad ::= \quad \langle class \rangle$$
$$| \quad \langle role \rangle$$
$$\langle class \rangle \quad ::= \quad \langle concept \rangle$$
$$| \quad \langle attribute\text{-}set \rangle$$
$$| \quad \langle number\text{-}set \rangle$$

In the simplest case a *class* is introduced as a *primitive concept* by stating necessary conditions which determine its membership.

    University :< Organization

A university is – necessarily – an organization.

Concept Terms

In order to produce composite descriptions out of such *primitive* introductions, *concepts* can be joined with other *concepts*, or their relations to other *concepts* can be restricted w.r.t. range and number of fillers.[2] If a composite description contains all necessary and sufficient conditions for an object to instantiate it the description is introduced as a *defined concept*.

$$\langle concept \rangle \quad ::= \quad \langle concept\text{-NAME} \rangle$$
$$| \quad \langle concept \rangle \text{ and } \langle concept \rangle$$
$$| \quad \textbf{all} \, (\, \langle role\text{-NAME} \rangle \, , \langle class \rangle \,)$$
$$| \quad \textbf{atleast} \, (\, \langle \text{NUMBER} \rangle \, , \langle role\text{-NAME} \rangle \,)$$
$$| \quad \textbf{atmost} \, (\, \langle \text{NUMBER} \rangle \, , \langle role\text{-NAME} \rangle \,)$$
$$| \quad \textbf{all1} \, (\, \langle role\text{-NAME} \rangle \, , \langle class \rangle \,)$$
$$| \quad \textbf{anything} \, | \, \textbf{nothing}$$

    Esprit-Eligible   :=   Company
                           AND ALL1(has-residence, CEC-country)

An Esprit-Eligible company is defined as a company which has its residence in a CEC-country.

The notion of forming a structured description, whose meaning is solely determined by its internal structure is a central characteristic for the terminological approach. It allows us to explicate all implicitly given subsumption relations between *classes*, and it allows us to deduce the correct class membership from the set of features known about an instantiating *object*. We will discuss these inferences, which are referred to as *classification* and *realization*, in more detail below.

---

[2]The number of role fillers can be given by specifying upper or lower bounds of a number interval. For roles having at least one filler we use the abbreviation all1.

Role Terms

Roles denote relations between concepts. In analogy to the *primitive concept* hierarchy they form a hierarchy of *primitive roles*.

$$\langle role \rangle \ ::= \ \langle role\text{-NAME} \rangle$$
$$| \ \langle role \rangle \ \textbf{and} \ \langle role \rangle$$
$$| \ \textbf{domain} \, (\, \langle concept \rangle \,)$$
$$| \ \textbf{range} \, (\, \langle class \rangle \,)$$

```
Consortium        :<   ANYTHING
has-members       :<   DOMAIN(Consortium)
                       AND RANGE(Organization)
```

Has-members is a relation between consortium and organization.

The global view of the *role* hierarchy which is implied by this kind of *role* introduction should be contrasted with the view of *roles* specifying certain local restrictions at *concepts* (we already made use of this above):

```
Esprit-Consortium :=   Consortium
                       AND ALL1(has-members, Esprit-Eligible)
```

An Esprit Consortium is defined as a consortium which has only Esprit-Eligible members.[3]

Local restrictions may not be in conflict with the global *role* hierarchy.

Attribute Set and Number Set Terms

While *concepts* and *roles* constitute the classical representational core of a term description language, some variants for representing *concepts* are added: *attribute-sets* are used for dealing with sets of attribute values in cases where a set can better be represented by enumerating all of its

---

[3]In fact, the actual definition of an Esprit-Consortium is:

Each Consortium must include at least two independent industrial organisations from the Community not established in the same Member State (cf. [14]).

Dealing with these kinds of additional constraints requires a language which supports more complex descriptional techniques for *roles*. Quantz provides the complete solution in [43].

elements.[4] The key word **attribute** denotes the set of all possible attribute values.

$$\langle attribute\text{-}set \rangle \quad ::= \quad \textbf{aset} \, (\, \langle attribute\text{-}\text{NAME} \rangle^{+} \,)$$
$$| \quad \textbf{attribute}$$

```
European-Country  :=  ASET(Belgium, Denmark, England,
                          France, Germany, Greece, Ireland, Italy,
                          Luxemburg, Netherlands, Portugal, Spain,
                          Bulgaria, Czechoslovakia, Finland,
                          Hungary, Yugoslavia, Norway, Austria,
                          Poland, Romania, Russia, Sweden,
                          Switzerland, Turkey)
CEC-Country       :=  ASET(Belgium .. Spain, European-Country)
```

In the abbreviated definition we take advantage of the total order of the elements given in the initial definition.

For a more convenient way of dealing with ranges of numbers *number-sets* are used. The key word **number** denotes the set of all integers.

$$\langle number\text{-}set \rangle \quad ::= \quad \langle \text{NUMBER} \rangle$$
$$| \quad < \langle \text{NUMBER} \rangle \, | \, > \langle \text{NUMBER} \rangle$$
$$| \quad \textbf{number}$$

```
sme := Company AND ALL1(has-employees, <50)
```

A small and medium enterprise is defined as a company with at most 50 employees.

In addition to the constructs described above most existing systems based on term description languages support a large variety of additional functionalities, e.g. for getting knowledge base scheme information, revising the scheme, dealing with simple set operations on attribute-sets and number-sets, and more.

Let us look at the example now in a more complete version:

---

[4]While this kind of definition is also known as *extensional definition* it should not be confused with the extensional level where we are dealing with instances.

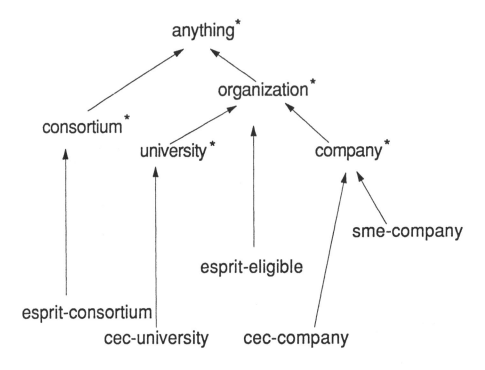

Figure 1: Concept Hierarchy (primitive concepts are indicated by an asterisk)

```
European-Country  := ASET(Belgium .. Turkey)
CEC-Country       := ASET(Belgium .. Spain, European-Country)
Organization      :< ANYTHING
has-employees     :< DOMAIN(Organization) AND RANGE(NUMBER)
has-residence     :< DOMAIN(Organization) AND RANGE(ATTRIBUTE)
has-name          :< DOMAIN(Organization) AND RANGE(ATTRIBUTE)
Consortium        :< ANYTHING
has-members       :< DOMAIN(Consortium) AND RANGE(Organization)
University        :< Organization
CEC-University    := University AND ALL1(has-residence, CEC-Country)
Company           :< Organization
CEC-company       := Company AND ALL1(has-residence, CEC-Country)
SME-company       := Company AND ALL1(has-employees, <50)
CEC-SME-company   := SME-company AND CEC-company
European-SME      := SME-company AND ALL1(has-residence, European-Country)
Esprit-Eligible   := Organization AND ALL1(has-residence, CEC-Country)
Esprit-Consortium := Consortium AND ALL1(has-members, Esprit-Eligible)
```

The sequence of *term equations* forms a *concept* and a *role* hierarchy (and also the trivial, but useful *attribute-set* and *number-set* hierarchies). Seen as a data structure the concept hierarchy forms a labelled directed acyclic graph which – for a slightly extended version of our example sequence – are shown in Figures 1 and 2.

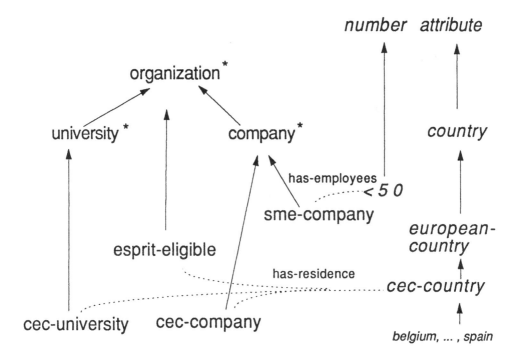

Figure 2: Concepts with Role Restrictions

2.2 Object Description and Retrieval

In the process of formalizing the domain for further processing, so far, we have a collected a number of *term equations* which constitute the *terminological model* of the domain. The next step is to introduce assertions about *objects* which instantiate the *classes* of the terminological model. In the simplest case an assertion is specified by introducing a *unique name* (it may also have synonyms) as identifier and specifying the *class* the *object* instantiates:

```
u-52  =    University
c-99  =    Company
```

A sequence of such *object equations* is called the *assertional knowledge base*. In addition, specifications may consist of complex assertional terms integrating composite terms of the term description language:

$$
\begin{array}{rcl}
\langle object\text{-}tell \rangle & ::= & \langle object\text{-NAME} \rangle = \langle class\text{-}expression \rangle \\
\langle object\text{-}ask \rangle & ::= & \langle object\text{-VAR} \rangle = \textbf{getall } \langle class\text{-}expression \rangle \\
\langle class\text{-}expression \rangle & ::= & \langle concept \rangle \\
& | & \langle attribute\text{-}set \rangle \\
& | & \langle concept \rangle \textbf{ with } \langle role \rangle : \langle value\text{-}expression \rangle \\
\langle value\text{-}expression \rangle & ::= & \langle object\text{-NAME} \rangle \\
& | & \textbf{close} (\, \langle value\text{-}expression \rangle \,) \\
& | & (\, \textbf{all } \langle class\text{-}expression \rangle \,) \\
& | & \langle value\text{-}expression \rangle \textbf{ and } \langle value\text{-}expression \rangle
\end{array}
$$

In fact, it is the salient point of the approach presented here that the knowledge base access language also allows the integration of descriptive parts of the intensional level. Starting with the terminological scheme given in our example we can enter information about instantiating objects, e.g. an object C-99:

C-99 *is a company.*

C-99 *has 30 employees.*

C-99 *has its residence in Italy.*

All these entries are taken in conjunction. In general, the object identifier may point to a *composite description* of the term description language without requiring any information about the *exact name* of the class this object ultimately belongs to. Determining the membership is done by the process of *realization* which can be seen as the counterpart of *classification* on the extensional level. Querying, e.g., for *all Esprit-Eligible SMEs* which are known so far in the knowledge base, we can expect that object C-99 will be retrieved which is achieved by *classification* where, in this case, it is explicated from the above definitions that *all European SME companies* are subsumed by the set of *Esprit-eligible companies*.

In addition, the language also shows descriptional means called *value expressions* by which information on the fillers of roles can be determined. If e.g. we use an **and** in the value expression that restricts a role, those objects are retrieved that have the fillers in conjunction as their role fillers.[5]

In general, an assertional language designed for practical usage should provide many additional features including operators for anonymous referencing, recursive nesting, chaining, and others.[6]

---

[5]By using such expressions without the **close** constructor partial knowledge about such role fillers can also be dealt with, i.e. the object retrieved in this case may have additional role fillers, too.

[6]Kindermann has designed and implemented such an extended language for the BACK system, as described in [42].

2.3   Characteristics of the Terminological Approach

Recalling the main principles of most systems designed under the terminological paradigm we should stress the features of a *well-defined semantics*,
the *specialized reasoning* style for various reasoning types, and the *class-
based* and *object-based organization* of domain entities.[7]

We should review how these essential elements influence two main aspects
of an information system, expressiveness characteristics and inferential capabilities.

Expressive Knowledge Base Language

As already pointed out, the term description language can be seen as a
language for the specification of the knowledge base scheme (in analogy to
a data definition language), and the language part for assertional *objects*
as a knowledge base assertion and query language (in analogy to a data
manipulation language). However, the knowledge base access language we
presented within the previous section focusses on a close integration of both
description levels, the intensional description language, and description of
extensional ground facts.[8] How would such a system measure up in a typical
database scenario?

In conventional database approaches it is necessary that all names be known
beforehand when any information service is asked for. For the task of
navigation or information retrieval most information on how to identify the
corresponding DB tables has to be already present. Or – in other words
– once you have to know all the names and you have to tell the system
exactly where to look for the information you hardly need the answer any
more. In contrast, the terminological approach is more flexible and easy-to-
use. Of course names are used here, too, namely the names for *classes* and
roles introduced as *primitives*. However, using the technique of building

---

[7]There is some confusion in the literature as to the appropriate expression here. We
use the expression *object-based* here as a metaphor for the system's epistemological characteristics. Often such systems are also called *object-centered, concept-based*, or *frame-based*
(sometimes the expression *frame-based* systems is used only for primitive object hierarchies).

[8]The idea of a uniform language differs from merely offering a number of single tools
and lower level access functions and network editing utilities. The language integration
aspect becomes significant in particular for knowledge base revision: It makes a difference
whether the system offers a number of modification and revision facilities leaving part of
the responsibility to the user, or if monotonic and non-monotonic update oprations are
integrated into the syntax and semantics of a uniform interface language.

definitions frees the user from the use of names to a large extent since the system is actually *reasoning* on descriptions, detecting equivalences, etc.

Moreover, dealing with descriptions in the way presented here is not only an advantage for knowledge base access by a user. In an extended scenario, we could imagine external system components accessing the representational service. A component collecting data and monitoring information from sensors or from permanently incoming messages would produce a vast number of descriptions. Since these descriptions would be generated automatically, the need for reasoning on descriptions becomes evident.[9]

The language expressiveness can also be exploited in improving conventional query answering techniques. Instead of producing only answers in the form of an extensional enumeration of the resulting set it is convenient to also incorporate intensional parts into the answer, thus providing a more dense type of query answering which is also able to deal with partial knowledge [8]. The semantically well-founded integration of the extensional and intensional levels within the terminological approach would seem to be a good starting point for more advanced answer generating.

Inferential Capabilities

Turning back to the initial, rather ambitious expectations about the range that knowledge representation is supposed to cover, the service offered by a terminological system may look somewhat disappointing. And even if one accepts the restriction that terminological systems focus mainly on representing *definitional* knowledge, the impression does not improve since even this service is not complete. Lacking any non-standard logical characteristics, the system seems to be able to deal only with *technical*, or *artificial* definitions, in contrast to *natural* definitions where often other descriptional methods are applied such as prototypes, defaults, or examples (cf. [18] for a detailed critique).

However, this impression is misleading. It is obvious that basing a system on a formal semantics and opting for a computationally tractable system first leads to a restriction to a well-known logics. In addition, the terminological reasoner is only conceived as a core component, and starting with a good standard maintenance of descriptions is a sound basis for extensions to be built on top of these descriptions. Let us review the core of the inferential service provided by terminological reasoning:

---

[9]For these applications – as for many others – the use of descriptions dependent on temporal relations is highly desirable. The integration of this type of reasoning within the terminological approach has been investigated in [48].

- *Consistency checking* is performed for all intensional descriptions. Relations between classes are *inherited* to all specializations of these classes. Consistency conflicts and incoherencies caused by combining complex descriptions are detected. Checking for valid descriptions can be exploited in the scheme definition phase as well when querying for instances, thus providing an efficient kind of *query validation*.

- *Classification* is the process of finding the correct place for composed terminological descriptions in the hierarchy. Together with *realization*, i.e. the process of finding the best description for asserted objects, classification forms the central inference of the reasoning process.[10]

- *Completion* of partial descriptions of objects can be performed if appropriate information is given at the intensional level (e.g. in the form of maximal number restrictions), or if *closing* expressions are used at the extensional level (stating that the specified values are the only values filling the specified roles).

All these inferences are strictly deductive and definitional. In addition, it is necessary to consider further techniques which state relationships which go beyond the purely definitional ones. For example, a statement such as

companies are disjoint from universities

means that a composition of both concepts (sme AND large-company) may be a proper definition on its own, however, there exists no instantiation by any object, a kind of disjointness which differs from definitional disjointness. Similarly, it is obvious that a rule such as

a company with at most 10 employees is also a company with at most 20 employees

should be treated differently from a rule such as

a SME-type company is also a dynamical company

---

[10] As we will see in the next section the ability to detect "forgotten" links in the scheme definition phase is only the minor contribution of classification. Its role is of more importance in the realization phase.

since the former contains purely definitional relations. In our example language these relationships can be expressed by the *disjointness* and *implication* constructs.

However, this integration of rules is sketched here only as a first step of extending the terminological approach – in this case by applying forward chaining rules to classes [8, 45]. Additional inferential modes have already been investigated, e.g. probabilistically weighted implication [36] and default techniques [30]. If such further modes are added a terminological reasoner can also be viewed as a core component of a classical rule-based reasoner: For production rules the *left-hand sides* of the rules are obviously equivalent to complex descriptions, thus being good candidates for *right-hand sides* of terminological term equations as presented above. Since the techniques for dealing with *right-hand sides* of these equations are well understood, we gain a solid structuring basis for complex interdependencies which also overcomes the limitations of the simple rule model [52].

Although experiences in various case studies[11] have – in principle – shown the usefulness of description and query techniques, a broader utilization of an expressive access language depends on such additional integration with other programming paradigms. Further work is therefore needed in extending the framework by adding non-deductive reasoning modes, thus finally exploiting the descriptive capabilities of a terminological language in an inferentially powerful information system.

3 Persistency Management by Semantic Indexing

Our presentation so far has focussed on the abstract logical level of terminological logics. In this section we will try to show how this logical structure can be exploited in information processing. A number of advantages should be immediately obvious. The *classification* inference, for instance, can be employed in the process of designing a database scheme by creating a partial ordering over the concepts – the concept hierarchy. A graphical depiction of the hierarchy can then be used to verify that the concepts end up in the right places. Additionally, it is possible to transform the schema into a *minimal* form, if desired [6].

---

[11]See e.g. [16] for a case study on using the BACK system for modelling the legal structure of a large company group, and [17] on using the ARGON system within a software information system.

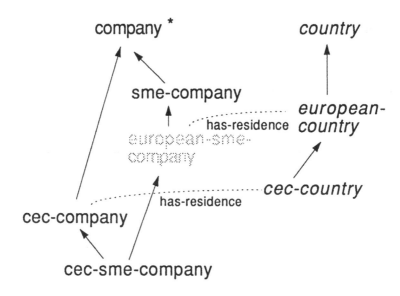

Figure 3: Extended Concept Hierarchy

Although such an application of terminological reasoning is certainly an advantage, it is not the only one and not the most important one. The more important application of classification in the database context is *query-processing* and *indexing*.

As we have seen above, queries are formulated in essentially the same language as the one used for defining the database schema, and so do not require navigation through the schema but are entirely declarative. Furthermore, it is possible to apply classification to queries as well. An obvious benefit one gains by classifying a query into the schema is that the query will be validated, i.e., it will be checked whether the concept used to express the query can possibly denote something. This is only the case if the query concept is different from the least concept, the empty concept called Nothing. For instance, if the query contained the concept

```
SME-companies AND ALL1(has-employees,>50),
```

it would be rejected because it is equivalent to Nothing.

Provided the query is semantically well-formed, i.e., not equivalent to Nothing, it is necessary to retrieve the objects that conforms to the concept description expressed in the query. The most simple-minded way to implement retrieval would be to scan all objects and to check whether they satisfy the concept description, i.e., whether they are an *instance* of the query-concept, and to return as the answer-set the set of all instances of the query-concept. Although this is conceptually correct, it is also quite inefficient.

In conventional database architectures, indexing is a technique to avoid scanning all instances in order to evaluate a query, but to select a small subset which is tested against the query. For instance, in the example in the last section, we would index with respect to has-name, has-residence, and has-employees, provided that we would like to access organizations by their names, residence, and their number of employees, respectively. Setting up such index structures is one of the important physical design considerations and can have dramatic effects on the performance of a database system. There are some disadvantages of this conventional technique, however. First, the indices must be set up once and for all and reorganization is costly and cumbersome. Second, the logical database schema is relatively neutral with regard to indexing. Which fields are to be indexed is mostly a pragmatic issue, although such information could come from information about how the user views the application domain.

Terminological logics offer a way to organize indexing along the "semantic" dimension. The idea is to set up the index parallel to the way the terminology is organized, i.e., for each concept (pointers to) all objects that are instances of this concept are stored.[12] Using this organization which exploits the "semantic" space of the terminology – and for this reason we call it *semantic indexing* – makes query processing simple and straightforward.

A query is processed in the same way as a new concept definition is handled, namely, the query-concept is temporarily classified into the concept hierarchy. As a result, either an *equivalent* concept or the set of *immediate subsumers* and *immediate subsumees* is returned. In the former case, the answer set is simply the set of all objects that are instances of the concept equivalent with the query-concept. In the latter case, the union of the sets of instances of the immediate subconcepts are included in the answer set and only objects that are instances of the immediate superconcept but not instances of the immediate subconcepts have to be checked against the query concept.

As an example, let us assume that we want to retrieve all European SME's, i.e., all instances of the concept

SME-company AND ALL1(has-residence, European-Country).

---

[12]How to implement such a structure on top of a relational database system is analyzed in [23, 24].

Classifying this concept into the hierarchy returns SME-company as the only immediate superconcept and CEC-SME-company as the only immediate subconcept (see the details depicted in Fig. 3). Using this result, all instances of CEC-SME-company are returned and all instances of SME-company that are not instances of CEC-SME-company are checked against the query-concept and included in the answer-set if they satisfy the condition.

Obviously, this technique of semantic indexing can also be used to reorganize the index structure dynamically in a transparent way. For instance, if (semantically equivalent) queries are posed very often, the system can introduce an anonymous concept on its own, extending the indexing structure in this way. Furthermore, the introduction of such anonymous concepts can also be done on information-theoretic grounds, i.e., the database can use self-organization techniques in order to tune its performance (see, e.g., [28]).

There is a price to pay for such an organization, however. Every time a new fact about an instance is entered into the database, the instance – and perhaps other related instances as well – have to be checked and the index structure has to be reorganized, a process often called *realization.*

As an example, assume we incrementally enter information about the company c-99 as informally described in the previous Section:

```
c-99  =    Company with has-employees:close(30)
c-99  =    Company with has-residence:close(Italy).
```

Basically, the first statement asserts that the company c-99 has 30 employees, and the second statement asserts that the company is located in Italy. The both close expressions are used to state that the specified values are the only values filling the specified roles of the described object. After the first statement c-99 will be recognized as a SME-company, after the second statement as a CEC-SME-company.

As a side-effect this process detects violations of integrity constraints expressed in concept-definitions, namely, when an object is recognized as instances of the "empty" concept Nothing. For example, when instead of the second statement above the following assertion was put into the database

```
c-99  =    CEC-Company with has-residence:Poland,
```

an inconsistency would be detected. However, asserting that c-99 is a Company *and* a University at the same time would not lead to an inconsistency since these concepts were not defined to be *disjoint* in the schema. In fact, a private university may be considered as both a university and a company.

4 Related Work

The approach described so far is not unique. A number of other research projects aim at similar goals and use comparable techniques. In the following, some of this work will be surveyed. Furthermore, a number of theoretical results concerning reasoning in terminological logics have been achieved recently which sometimes apply directly to work done in the related area of semantic data models and complex object data models.

4.1 Terminological Representation and Reasoning Systems

Besides focussing on the development of a number of terminological knowledge representation systems which were built to support AI applications,[13] some research groups explicitly address the problem of supporting database management with terminological reasoning techniques, for instance, the BACK system described in the previous sections (see also [51, 42, 44]).

The first such approach was probably the RABBIT system [50], which supports the user in selecting an appropriate restaurant using a technique called *query by reformulation*. This system was implemented on top of KL-ONE and used classification as a means for selecting the appropriate concepts in the concept network. It should be noted that this system did not differentiate between concepts and instances (or schema and data, respectively), but represented concepts and instances in a uniform way.

Subsequently, an information retrieval system called ARGON [40] was developed using the concepts and ideas of the RABBIT system. In contrast to RABBIT, however, in ARGON there is a clear-cut distinction between concepts and instances, and, more importantly, the system uses the technique of *semantic indexing* described above, which is implemented as part of the underlying representation system KANDOR [37]. ARGON was tested successfully on a small database of AI researchers (1500 individuals), on a TTL chip catalog [15],[14] and a knowledge-based software information system [17].

A more recent development is the CLASSIC system [8, 10], which takes into account recently achieved theoretical results concerning expressivity and tractability of terminological representation formalisms. In particular, the

---

[13]For instance, NIKL [21], KRYPTON [11], LOOM [30], QUIRK [7], SB-ONE [25]. For the state of the art and future developments, see [41].

[14]An interesting aspect of the TTL catalog application is that queries are always interpreted intensionally according to different levels of abstractions.

terminological logic used incorporates *co-reference constraints* – a quite powerful construction also used in similar formalisms such as ψ-terms [2] and *feature logic* [35]. This system is also aimed more at database applications because it supports updates of instances and means of applying (forward-chaining) rules to database entities. These rules do not act only as integrity constraints but they are also used to conclude additional information about database entities.[15]

4.2 Semantic Data Models and Complex Object Data Models

In the past decade it has become common knowledge in the database research field that traditional record-based database systems have many serious limitations (see, e.g., [22]), and a number of so-called "semantic" data models have been proposed in order to overcome these limitation by providing mechanisms and constructs that allow modelling of the kinds of relationships that naturally occur in an application. In a certain sense, the recent wave of object-oriented database models can be understood as a natural extension of this approach.

Semantic data models and object-oriented databases have in common the phenomenon that complex descriptions are used to describe types and/or classes and, consequently, reasoning about these descriptions became a research topic. Some of the approaches in this direction are directly influenced by work done in the area of terminological reasoning. For instance, terminological reasoning has been exploited in the design of entity-relationship schemata [5]. As another example, the semantic data model CANDIDE [4] employs terminological representation and reasoning techniques on all levels, namely, for schema design, query processing, and indexing – which is not surprising since CANDIDE is based on the representation system KANDOR mentioned above.

Independently from terminological logics, research in semantic data models and object-oriented database models started to analyze the structure of types induced by their defining descriptions for the purpose of *type inference* and *type checking*. One example is IFO [1], where an *dominance* ordering over "derived types" is computed. Another example is the object-oriented data model O_2 [26, 27]. O_2 permits the composition of class descriptions using some basic types, such as integers and strings, a tuple constructor, a set constructor, and class disjunctions. The semantics is based on set

---

[15]Basically, these rules are similar to the implicational rules described in Section 2.3.

theory similar to the semantics used in the area of terminological logics. Furthermore, a *refinement* ordering is defined over classes quite similar to the *subsumption* ordering discussed above. Although it is not possible to reconstruct O_2 in the terminological logic described here, more powerful language containing concept disjunction (see, e.g., [20]) could be used for this purpose. As pointed out above, however, although these database models might look similar, and indeed use similar algorithms, the intention for computing subsumption orderings is different in these approaches. It is not used to support query-evaluation or integrity control as we have sketched it in the previous sections, but only to do type checking.

One approach employing the technique of query-evaluation by classification is a non-standard database application for the management of chemical structures [28]. Here subgraph-isomorphism induces a partial ordering on labeled graphs, and this partial ordering is used to efficiently store and retrieve such graphs. The difference to the technique of semantic indexing described above is that there is no distinction between concepts and instances. Query-graphs are simply classified into the partial ordering and the immediately preceding and immediately succeeding graphs are considered as the answer.

4.3   Theoretical Issues

After having seen what can be done with terminological logics, one may pose the question of what algorithms do look like, how efficient they are, and in which way this depends on the expressiveness of the terminological logic chosen. In order to answer these questions, it is necessary to specify a rigorous and unambiguous semantics for the logics. The presentation of terminological logics in the first part of the paper has been deliberately informal. However, an intuitive and plausible Tarski-style semantics following the ideas first spelled out in [9] can be straightforwardly specified and is provided in the appendix. Based on that, the complexity of terminological reasoning has been investigated by a number of researchers and it has turned out that depending on the expressiveness of the terminological logic, the complexity of subsumption determination ranges from polynomial-time to undecidability. We will not go into the details, but only refer to some of the important results.

First of all, the ability to define concepts introduces a perhaps surprising source of complexity. To see this, one should note that in order to compute subsumption between two concepts the definitions of these concepts is usually "expanded" before the concepts are compared, i.e., all defined terms are substituted by their definitions until the expressions do not contain a defined term any more. This can be only done if there are no *cycles* in the terminology, i.e., if the defined terms do not refer directly or indirectly to themselves.

This expansion can lead to an exponential blow up in the worst case, however. Worse yet, most probably (under the assumption that $P \neq NP$) there is no algorithm that computes subsumption between two concepts defined in a terminology in polynomial time, provided the terminological logic contains at least the operators AND and ALL [34]. This is shown by reducing equivalence of cycle-free nondeterministic finite automatons to subsumption determination. The same proof technique can be used to show that this complexity result also holds for terminological logics that contain concept disjunction (OR) instead of concept conjunction (AND), and by that it follows that the determination of the *refinement* ordering in O_2 as described in [27] is NP-hard as well – a fact which apparently was not noticed by the authors.[16] The polynomial algorithm for computing *refinement* in [27] turns out to be incomplete. One example of unnoticed refinement in O_2 is the following. Assume three incomparable classes X, Y, Z, and the following declarations:

```
A    =    X OR Y
B    =    Y OR Z
C    =    X OR Y OR Z
```

A OR B is obviously a subtype, a refinement, of C. However, the algorithm in [27] does not notice that.[17]

In any case, the apparent intractability of the subsumption problem in terminologies seems not to be relevant from a practical point of view. In almost all cases subsumption can be determined efficiently. Indeed it is possible to specify some reasonable restrictions on the form of a terminology that permit a provably polynomial algorithm [34].

Second, there might be the question of what happens if we drop the restriction mentioned above that terminologies have to be cycle-free. As it turns out, this is possible without giving up too much. If the terminological logic is intended to be used for *defining* concepts, a fixpoint – in this case a greatest fixpoint – semantics is the appropriate means to formalize the

---

[16]As a matter of fact, the the *refinement* problem in O_2 is PSPACE-complete since in O_2 cycles are permitted, so the results cited below apply.

[17]Although disappointing, this result does not affect the O_2 system as it exists. As pointed out to us by C. Lecluse, disjunctive classes are not part of the implemented system, but they were only considered as one possible extension of the system.

meaning of cyclic concepts (see [33, 3]).[18] The complexity result mentioned above leads to PSPACE-hardness in this case.

Third, ignoring the complexity introduced by defining concepts, one may ask how difficult subsumption determination is if the concepts are already "expanded". Obviously, this depends on the expressiveness of the term description language. As first shown in [9] and [31], only very simple term description languages permit a polynomial subsumption algorithm. Subsequently, researchers concentrated on developing ways to circumvent this complexity trap by radically restricting the expressiveness (as exercised, e.g., in KRYPTON [11], and CLASSIC [8]), by using weaker semantics that permit tractable subsumption determination [38], or by restricting the inferential capabilities in a pragmatic way leading to *sound* and fast but *incomplete* systems, such as BACK [31] and LOOM [29]). Only recently, complete algorithms for a number of powerful term description languages for which the subsumption problem is NP-hard have been developed [47, 20]. It is yet unclear, however, whether these algorithms are feasible, i.e., whether the worst case does not show up for "naturally" occurring concepts.

Interestingly, terminological logics seem to be decidable in most cases. There seems to be only one construction which causes trouble, namely, so-called *role-value-maps*, also called *co-reference constraints*, which allow one to state that the role-fillers of two "role-chains" are identical.[19] However, if one constrains the interpretation of roles used in co-reference constraints to be single-valued, subsumption becomes decidable again, a fact exploited in CLASSIC.[20] This holds only as long as there are no cycles in the terminology, though. Co-reference constraints over single-valued roles *plus* cycles in the terminology are again undecidable with respect to subsumption [33]. Unfortunately, using the same argument, it can be shown that *consistency* of a CLASSIC database is in general undecidable. Technically speaking, it is undecidable whether there exists a consistent completion of a CLASSIC database.[21]

Accepting the complexity of subsumption, there remains the question of how efficient classification and the associated technique of semantic index-

---

[18]For natural language applications other options may be more adequate, however.

[19]For instance, if we want to define a subconcept of **Father** such that his **last-name** is identical with the **last-name** of his **children**, a role-value-map has to be used. In general, such constructions lead to undecidability [39, 46].

[20]As a matter of fact, co-reference constraints over single-valued roles are computationally very well-behaved, which leads to a polynomial subsumption algorithm in CLASSIC – an insight gained from the analyses of so-called feature logics (see, e.g., [49]).

[21]It is an open problem whether consistency in a terminological logic containing no co-reference constraints is decidable if rules are added to the terminological logic.

ing may be. Unfortunately, there is no straightforward answer. Only very little is known about the average complexity of inserting new elements into a partial order – a fact which may seem surprising in face of how much is known about total orders and about trees. It is obvious that semantic indexing as described in Section 3 is more efficient than sequentially scanning all facts. However, it is not clear how much efficiency one gains or loses compared with conventional indexing techniques. Nevertheless, there is some empirical evidence that using partial orders instead of indices over simple properties speeds up access considerably – at least in the case of the nonstandard databases for chemical structures [28].

Finally, there is the question of what price we have to pay for semantic indexing at the time when instances are entered into the database. Again this depends on the expressiveness of the terminological logic used. In general, *realization*, i.e., computing the classes an object is an instance of, is not easier than subsumption. As a matter of fact, in a standard *open world* database, subsumption can be reduced to realization. On the other hand, realization is usually not much harder than subsumption since from a theoretical point of view a slight modification of the subsumption algorithm can be used [19]. For *closed world* databases the picture is sometimes different, however [32, Chap. 4].

5 Conclusion

Representation and reasoning with terminological logics, a subfield of knowledge representation, appears to have reached a point where the techniques developed in this field seem to be applicable to databases and information system. First of all, these logics support the representation of complex relationships, a necessity for any database model that intends to go beyond record structures. Second, reasoning in these logics can be exploited for various important tasks in database and information management, such as schema design, query evaluation and indexing. In particular, the last point seems to be very interesting since indexing can be done in a way which parallels the semantic structure of the schema, a reason for calling this technique *semantic indexing*. Third, the theoretical foundations of terminological logics have been analyzed to a depth that makes the results applicable to other areas in computer science. In particular, we were able to identify a problem in the object-oriented database model O_2 by applying results from the analyses of terminological logics. Although this sounds quite promising, it will probably take some time before terminological logics will be applied in database and information systems. There are a number of important theoretical and practical problems which have to be solved first. For instance, the problem of finding efficient physical design methods for systems based on terminological logics has not been tackled yet. Also there is the problem of designing *complete* (in some sense) query languages.

Summarizing, although we do not have an instant solution for all problems appearing in the area of database and information management, we hope to have shown that the paradigm of terminological logics has something to offer to future information systems.

Acknowledgement

We are grateful to Sonia Bergamaschi and the colleagues of our project groups for helpful comments on earlier drafts.

Appendix A: Syntax

KB Terminological Schema (TBox)

$$
\begin{array}{lll}
\langle term\text{-}tell\rangle & ::= & \langle concept\text{-}\mathrm{NAME}\rangle & :< \langle concept\rangle \\
& | & \langle concept\text{-}\mathrm{NAME}\rangle & := \langle concept\rangle \\
& | & \langle attribute\text{-}set\text{-}\mathrm{NAME}\rangle & := \langle attribute\text{-}set\rangle \\
& | & \langle role\text{-}\mathrm{NAME}\rangle & :< \langle role\rangle \\
& | & \mathbf{disjoint}(\langle concept\rangle,\langle concept\rangle) \\
& | & \mathbf{implies}(\langle concept\rangle,\langle concept\rangle)
\end{array}
$$

$$
\begin{array}{lll}
\langle term\text{-}ask\rangle & ::= & \mathbf{subsumes}(\langle class\rangle,\langle class\rangle)
\end{array}
$$

KB Objects (ABox)

$$
\begin{array}{lll}
\langle object\text{-}tell\rangle & ::= & \langle object\text{-}\mathrm{NAME}\rangle = \langle class\text{-}expression\rangle \\
\langle object\text{-}ask\rangle & ::= & \langle object\text{-}\mathrm{VAR}\rangle \ = \mathbf{getall}\ \langle class\text{-}expression\rangle
\end{array}
$$

$$
\begin{array}{lll}
\langle class\text{-}expression\rangle & ::= & \langle concept\rangle \\
& | & \langle attribute\text{-}set\rangle \\
& | & \langle concept\rangle\ \mathbf{with}\ \langle role\rangle : \langle value\text{-}expression\rangle \\
\langle value\text{-}expression\rangle & ::= & \langle object\text{-}\mathrm{NAME}\rangle \\
& | & \mathbf{close}(\langle value\text{-}expression\rangle) \\
& | & (\mathbf{all}\ \langle class\text{-}expression\rangle) \\
& | & \langle value\text{-}expression\rangle\ \mathbf{and}\ \langle value\text{-}expression\rangle
\end{array}
$$

Term Description Language (TDL)

$$
\begin{array}{rcl}
\langle term \rangle & ::= & \langle class \rangle \\
& | & \langle role \rangle \\
\langle class \rangle & ::= & \langle concept \rangle \\
& | & \langle attribute\text{-}set \rangle \\
& | & \langle number\text{-}set \rangle
\end{array}
$$

$$
\begin{array}{rcl}
\langle concept \rangle & ::= & \langle concept\text{-NAME} \rangle \\
& | & \langle concept \rangle \ \textbf{and} \ \langle concept \rangle \\
& | & \textbf{all}(\langle role\text{-NAME} \rangle, \langle class \rangle)) \\
& | & \textbf{atleast}(\langle \text{NUMBER} \rangle, \langle role\text{-NAME} \rangle)) \\
& | & \textbf{atmost}(\langle \text{NUMBER} \rangle, \langle role\text{-NAME} \rangle)) \\
& | & \textbf{all1}(\langle role\text{-NAME} \rangle, \langle class \rangle)) \\
& | & \textbf{anything} \,|\, \textbf{nothing}
\end{array}
$$

$$
\begin{array}{rcl}
\langle role \rangle & ::= & \langle role\text{-NAME} \rangle \\
& | & \langle role \rangle \ \textbf{and} \ \langle role \rangle \\
& | & \textbf{domain}(\langle concept \rangle) \\
& | & \textbf{range}(\langle class \rangle)
\end{array}
$$

$$
\begin{array}{rcl}
\langle attribute\text{-}set \rangle & ::= & \textbf{aset}((\langle attribute\text{-NAME} \rangle)^+) \\
& | & \textbf{attribute}
\end{array}
$$

$$
\begin{array}{rcl}
\langle number\text{-}set \rangle & ::= & \langle \text{NUMBER} \rangle \\
& | & \textbf{<}\langle \text{NUMBER} \rangle \,|\, \textbf{>}\langle \text{NUMBER} \rangle \\
& | & \textbf{number}
\end{array}
$$

Non-Terminals containing the substrings NAME, NUMBER, or VAR indicate terminal names, numbers, or variables of the host language.

Appendix B: Semantics

In the following, we specify the set-theoretic semantics of the TDL. The symbols v, v_i are used for class terms, c, c_i for concept terms, a, a_i for attribute set terms, t, t_i for attribute value terms, n, n_i for number set terms, p, p_i for number terms, and r, r_i for role terms. $|S|$ denotes the cardinality of the set S. d, e are used to denote elements of V, and i denotes an element of I.

Definition 1 *Let D, A, and I be mutually disjoint sets, where D, the domain, is some arbitrary set, A is the set of attribute values, and I is the set of all integers. Let V be the union of D, A, and I. An* **extension function** $\mathcal{E}$ *is a function* *mapping classes to subsets of V, concepts to subsets of D, attribute sets to subsets of A, attribute values to elements of A, number sets to subsets of I, numbers to elements of I, and roles to subsets of $D \times V$, such that for all nonnegative integers m:*

$$\mathcal{E}[\text{anything}] = D$$
$$\mathcal{E}[\text{nothing}] = \emptyset$$
$$\mathcal{E}[c_1 \text{ and } c_2] = \mathcal{E}[c_1] \cap \mathcal{E}[c_2]$$
$$\mathcal{E}[\text{all}(r, v)] = \left\{ d \in D : \forall \langle d, e \rangle \in \mathcal{E}[r] \Rightarrow e \in \mathcal{E}[v] \right\}$$
$$\mathcal{E}[\text{atleast}(m, r)] = \left\{ d \in D : |\{e : \langle d, e \rangle \in \mathcal{E}[r]\}| \geq m \right\}$$
$$\mathcal{E}[\text{atmost}(m, r)] = \left\{ d \in D : |\{e : \langle d, e \rangle \in \mathcal{E}[r]\}| \leq m \right\}$$
$$\mathcal{E}[\text{all1}(r, v)] = \mathcal{E}[\text{all}(r, v)] \cap \mathcal{E}[\text{atleast}(1, r)]$$

$$\mathcal{E}[r_1 \text{ and } r_2] = \mathcal{E}[r_1] \cap \mathcal{E}[r_2]$$
$$\mathcal{E}[\text{domain}(c)] = \mathcal{E}[c] \times V$$
$$\mathcal{E}[\text{range}(v)] = D \times \mathcal{E}[v]$$

$$\mathcal{E}[\text{attribute}] = A$$
$$\mathcal{E}[\text{aset}(t_1 \ldots t_n)] = \{\mathcal{E}[t_i] : 1 \leq i \leq n\}$$

$$\mathcal{E}[\text{number}] = I$$
$$\mathcal{E}[> p] = \left\{ i \in I : i > \mathcal{E}[p] \right\}$$
$$\mathcal{E}[< p] = \left\{ i \in I : i < \mathcal{E}[p] \right\}$$

Definition 2 *Let $\mathcal{E}$ be any extension function and τ any term-introduction or restriction. Let x_1 be a name and x_2 be some term and let c_1 and c_2 be two concept names that were introduced by $c_i :< \ldots$. Then $\mathcal{E}$ satisfies τ iff*

$$\mathcal{E}[x_1] = \mathcal{E}[x_2] \quad \text{if } \tau \text{ is of the form} \quad x_1 := x_2$$
$$\mathcal{E}[x_1] \subseteq \mathcal{E}[x_2] \quad \text{if } \tau \text{ is of the form} \quad x_1 :< x_2$$
$$\mathcal{E}[c_1] \cap \mathcal{E}[c_2] = \emptyset \quad \text{if } \tau \text{ is of the form} \quad \mathbf{disjoint}(c_1, c_2)$$

$\mathcal{E}$ satisfies a set of term-introductions or restrictions Θ iff $\mathcal{E}$ satisfies all elements of Θ.

Definition 3 *Let Θ be a set of term introductions and restrictions. Let x_1 and x_2 be two terms of the same syntactic category. Then x_1 is **subsumed by** x_2 iff for all $\mathcal{E}$ that satisfy Θ:*

$$\mathcal{E}[x_1] \subseteq \mathcal{E}[x_2]$$

References

[1] S. Abiteboul and R. Hull. IFO: A formal semantic database model. *ACM Transactions on Database Systems*, 12(3):525–565, Dec. 1987.

[2] H. Aït-Kaci. *A Lattice-Theoretic Approach to Computations Based on a Calculus of Partially Ordered Type Structures*. PhD thesis, University of Pennsylvenia, Philadelphia, Pa., 1984.

[3] F. Baader. Terminological cycles in KL-ONE-based KR-languages. In *Proceedings of the 8th National Conference of the American Association for Artificial Intelligence*, Boston, Mass., 1990.

[4] H. W. Beck, S. K. Gala, and S. B. Navathe. Classification as a query processing technique in the CANDIDE semantic data model. In *Proceedings of the International Data Engineering Conference, IEEE*, pages 572–581, Los Angeles, Cal., Feb. 1989.

[5] S. Bergamaschi, L. Cavedoni, C. Sartori, and P. Tiberio. On taxonomic reasoning in E/R environments. In C. Batini, editor, *Proceedings of the 7th International Conference on Entity Relationship Approach*, pages 301–312, 1988.

[6] S. Bergamaschi, C. Sartori, and P. Tiberio. On taxonomic reasoning in conceptual design. Technical Report 68, CIOC-CNR, Bologna, Italy, Mar. 1990.

[7] H. Bergmann and M. Gerlach. QUIRK: Implementierung einer TBox zur Repräsentation von begrifflichem Wissen. WISBER Memo 11, 2nd ed., Project WISBER, Department of Computer Science, Universität Hamburg, Hamburg, Germany, June 1987.

[8] A. Borgida, R. J. Brachman, D. L. McGuinness, and L. A. Resnick. CLASSIC: a structural data model for objects. In *Proceedings of the 1989 ACM SIGMOD International Conference on Mangement of Data*, pages 59–67, Portland, Oreg., June 1989.

[9] R. J. Brachman and H. J. Levesque. The tractability of subsumption in frame-based description languages. In *Proceedings of the 4th National Conference of the American Association for Artificial Intelligence*, pages 34–37, Austin, Tex., Aug. 1984.

[10] R. J. Brachman, D. L. McGuinness, P. F. Patel-Schneider, L. A. Resnick, and A. Borgida. Living with CLASSIC: When and how to use

a KL-ONE-like language. In J. Sowa, editor, *Principles of Semantic Networks*. Morgan Kaufmann, Los Altos, Cal., 1990. To appear.

[11] R. J. Brachman, V. Pigman Gilbert, and H. J. Levesque. An essential hybrid reasoning system: Knowledge and symbol level accounts in KRYPTON. In *Proceedings of the 9th International Joint Conference on Artificial Intelligence*, pages 532–539, Los Angeles, Cal., Aug. 1985.

[12] R. J. Brachman and J. G. Schmolze. An overview of the KL-ONE knowledge representation system. *Cognitive Science*, 9(2):171–216, Apr. 1985.

[13] M. L. Brodie. Future Intelligent Information Systems: AI and Database Technologies Working Together. In M. L. Brodie and J. Mylopoulos, editors, *Readings in Artificial Intelligence and Databases*. Morgan Kaufmann, San Mateo, Cal., 1988.

[14] Commission of the European Communities. Information Package for the Submission of Proposals to the European Strategic Programme for Research and Develpoment in Information Technology. Internal Information, DG XIII-A2, September 1989.

[15] F. Corella. Semantic retrieval and levels of abstraction. In L. Kerschberg, editor, *Expert Database Systems—Proceedings From the 1st International Workshop*, pages 91–114. Benjamin/Cummings, Menlo Park, Cal., 1986.

[16] M. Damiani, S. Bottarelli, M. Migliorati, and C. Peltason. Terminological Information Management in ADKMS. In *ESPRIT '90 Conference Proceedings*, Dordrecht, The Netherlands, 1990. Kluwer Academic Publishers. To appear.

[17] P. Devanbu, P. G. Selfridge, B. W. Ballard, and R. J. Brachman. A knowledge-based software information system. In *Proceedings of the 11th International Joint Conference on Artificial Intelligence*, pages 110–115, Detroit, Mich., Aug. 1989.

[18] J. Doyle and R. S. Patil. Two dogmas of knowledge representation: Language restrictions, taxonomic classifications, and the utility of representation services. Technical Memo MIT/LCS/TM-387.b, Laboratory for Computer Science, Massachusetts Institute of Technology, Cambridge, Mass., Sept. 1989.

[19] B. Hollunder. Hybrid inferences in KL-ONE-based knowledge representation systems. In H. Marburger, editor, *GWAI-90. 14th German Workshop on Artificial Intelligence*. Springer-Verlag, Berlin, Germany, 1990.

[20] B. Hollunder and W. Nutt. Subsumption algorithms for concept description languages. DFKI Report RR-90-04, German Research Center for Artificial Intelligence (DFKI), Kaiserslautern, Germany, 1990.

[21] T. S. Kaczmarek, R. Bates, and G. Robins. Recent developments in NIKL. In *Proceedings of the 5th National Conference of the American Association for Artificial Intelligence*, pages 978–987, Philadelphia, Pa., Aug. 1986.

[22] W. Kent. Limitations of record-based information models. *ACM Transactions on Database Systems*, 4(1):107–131, 1979.

[23] C. Kindermann. Class instances in a terminological framework – an experience report. In H. Marburger, editor, *GWAI-90. 14th German Workshop on Artificial Intelligence*. Springer-Verlag, Berlin, Germany, 1990.

[24] C. Kindermann and P. Randi. Object Recognition and Retrieval in the BACK System. KIT Report 86, Department of Computer Science, Technische Universität Berlin, Germany, To Appear.

[25] A. Kobsa. The SB-ONE knowledge representation workbench. In *Preprints of the Workshop on Formal Aspects of Semantic Networks*, Two Harbors, Cal., Feb. 1989.

[26] C. Lécluse, P. Richard, and F. Velez. O₂, an object-oriented data model. In *Proceedings of the 1988 ACM SIGMOD International Conference on Mangement of Data*, pages 424–433, Chicago, Ill., 1988.

[27] C. Lécluse, P. Richard, and F. Velez. Modelling complex structures in object-oriented databases. In *Proceedings of the 8th ACM SIGACT-SIGMOD-SIGART Symposium on Principles of Database-Systems*, pages 360–367, Mar. 1989.

[28] R. A. Levinson. A self-organizing pattern retrieval system for graphs. In *Proceedings of the 4th National Conference of the American Association for Artificial Intelligence*, pages 203–206, Austin, Tex., Aug. 1984.

[29] R. MacGregor. A deductive pattern matcher. In *Proceedings of the 7th National Conference of the American Association for Artificial Intelligence*, pages 403–408, Saint Paul, Minn., Aug. 1988.

[30] R. MacGregor and R. Bates. The Loom knowledge representation language. Technical Report ISI/RS-87-188, University of Southern California, Information Science Institute, Marina del Rey, Cal., 1987.

[31] B. Nebel. Computational complexity of terminological reasoning in BACK. *Artificial Intelligence*, 34(3):371–383, Apr. 1988.

[32] B. Nebel. *Reasoning and Revision in Hybrid Representation Systems*, volume 422 of *Lecture Notes in Computer Science*. Springer-Verlag, Berlin, Germany, 1990.

[33] B. Nebel. Terminological cycles: Semantics and computational properties. In J. Sowa, editor, *Principles of Semantic Networks*. Morgan Kaufmann, San Mateo, Cal., 1990. To appear.

[34] B. Nebel. Terminological reasoning is inherently intractable. *Artificial Intelligence*, 43:235–249, 1990. A preliminary version is available as IWBS Report 82, IBM Germany Scientific Center, IWBS, Stuttgart, Germany, October 1989.

[35] B. Nebel and G. Smolka. Representation and reasoning with attributive descriptions. In K.-H. Bläsius, U. Hedtstück, and C.-R. Rollinger, editors, *Sorts and Types in Artificial Intelligence*. Springer-Verlag, Berlin, Germany, 1990. To appear. Also available as IWBS Report 81, IBM Germany Scientific Center, IWBS, Stuttgart, Germany, September 1989.

[36] J. Heinsohn and B. Owsnicki-Klewe. Probabilistic inheritance and reasoning in hybrid knowledge representation systems. In W. Hoeppner, editor, *GWAI-88. 12th German Workshop on Artificial Intelligence*, pages 51–60. Springer-Verlag, Berlin, West Germany, 1988.

[37] P. F. Patel-Schneider. Small can be beautiful in knowledge representation. In *Proceedings of the IEEE Workshop on Principles of Knowledge-Based Systems*, pages 11–16, Denver, Colo., 1984. An extended version including a KANDOR system description is available as AI Technical Report No. 37, Palo Alto, Cal., Schlumberger Palo Alto Research, October 1984.

[38] P. F. Patel-Schneider. A four-valued semantics for terminological logics. *Artificial Intelligence*, 38(3):319–351, Apr. 1989.

[39] P. F. Patel-Schneider. Undecidability of subsumption in NIKL. *Artificial Intelligence*, 39(2):263–272, June 1989.

[40] P. F. Patel-Schneider, R. J. Brachman, and H. J. Levesque. ARGON: Knowledge representation meets information retrieval. In *Proceedings of the 1st Conference on Artificial Intelligence Applications*, pages 280–286, Denver, Col., 1984.

[41] P. F. Patel-Schneider, B. Owsnicki-Klewe, A. Kobsa, N. Guarino, R. MacGregor, W. S. Mark, D. McGuinness, B. Nebel, A. Schmiedel, and J. Yen. Term subsumption languages in knowledge representation. *The AI Magazine*, 11(2):16–23, 1990.

[42] C. Peltason, A. Schmiedel, C. Kindermann, and J. Quantz. The BACK system revisited. KIT Report 75, Department of Computer Science, Technische Universität Berlin, Germany, Sept. 1989.

[43] J. Quantz. Modelling and Reasoning with Defined Roles in BACK. The BACK system revisited. KIT Report 84, Department of Computer Science, Technische Universität Berlin, Germany, To Appear.

[44] J. Quantz and C. Kindermann. Implementation of the BACK System Version 4. KIT Report 78, Department of Computer Science, Technische Universität Berlin, Germany, Sept. 1990.

[45] K. Schild. Towards a theory of frames and rules. KIT Report 76, Department of Computer Science, Technische Universität Berlin, Germany, Dec. 1989.

[46] M. Schmidt-Schauß. Subsumption in KL-ONE is undecidable. In R. J. Brachman, H. J. Levesque, and R. Reiter, editors, *Proceedings of the 1st International Conference on Principles of Knowledge Representation and Reasoning*, pages 421–431, Toronto, Ont., May 1989.

[47] M. Schmidt-Schauß and G. Smolka. Attributive concept descriptions with unions and complements. *Artificial Intelligence*, 1990. To appear. A preliminary version of this paper is available as IWBS Report 68, IBM Germany Scientific Center, IWBS, Stuttgart, Germany, June 1989.

[48] A. Schmiedel. A temporal terminological logic. In *Proceedings of the 8th National Conference of the American Association for Artificial Intelligence*, pages 640–645, Boston, Mass., 1990.

[49] G. Smolka. A feature logic with subsorts. *Journal of Logic Programming*, 1990. To appear. Also available as LILOG Report 33, IBM Germany Scientific Center, IWBS, Stuttgart, Germany, May 1988.

[50] F. N. Tou, M. D. Williams, R. E. Fikes, A. Henderson, and T. Malone. RABBIT: An intelligent database assistant. In *Proceedings of the 2nd National Conference of the American Association for Artificial Intelligence*, pages 314–318, Pittsburgh, Pa., Aug. 1982.

[51] K. von Luck, B. Nebel, C. Peltason, and A. Schmiedel. The anatomy of the BACK system. KIT Report 41, Department of Computer Science, Technische Universität Berlin, Germany, Jan. 1987.

[52] J. Yen. A Principled Approach to Reasoning about the Specificity of Rules. In *Proceedings of the 9th National Conference of the American Association for Artificial Intelligence*, pages 701–707, Boston, Mass., 1990.

Conceptual Modeling of Database Applications

Gunter Saake

Informatik, Abt. Datenbanken
Technische Universität Braunschweig
D–3300 Braunschweig, F.R.G.

Abstract

The conceptual model is a high–level description of the functionality of a software system usually notated in a logic–based formalism. We discuss the special properties of descriptions of database applications using an appropriate conceptual formalism. A multi–layered approach for conceptual modeling of database applications is discussed together with first ideas on a design methodology using this approach. The specification of values, persistent database objects, temporal object evolution, update operations and arbitrary application processes is separated into different specification layers. Special attention is paid to problems of consistency checking for conceptual models and object–oriented modularization of design documents.

1 Introduction

Like in classical software design, in database design several design documents form a sequence starting from rather informal requirements and leading to the final implementation [ElN89]. The database design process may follow this sequence resulting in a software life cycle model of database design, or the single design documents may serve as different abstraction levels in an arbitrary software design process model. Usually, we distinguish the following documents or design phases both in database and software design:

- The *requirements analysis* collects the informal requirements in terms of the desired functionality of the system.

- The *conceptual model* is the first formal model of the designed system [BMS84]. It is formulated using a high–level formalism allowing abstract descriptions of the functionality without having to specify implementation details.

  In database design, the conceptual design phase results in a schema of a *conceptual data model* determining an abstract description of the database structures to be implemented together with a collection of transaction specifications [ElN89, Lip89]. In software design, the result is a formal specification of the functions realized by the implemented system.

- Based on the conceptual model, the system specification is refined leading to a detailed description in a formalism close to the concepts of the implementation framework. In database design, the result is the *logical schema* of the database.

- The last design phase is the final system *implementation*.

Following this sequence of design phases, we can identify the special role of conceptual design in the complete design process : The conceptual model is the first design document having a formal semantics, i.e., it is the document where all later documents have to be verified against. The quality of a conceptual model has therefore a significant influence on the later design phases because it is the basic reference model for the final system implementation. On the other hand, it is the 'contract' between the system developers usually familiar with formal techniques and the system users stating the informal requirements for system development.

There are well established methods and languages for the conceptual design of databases and of software functions. However, there is a lack of integrated conceptual design methodologies for designing *database applications*, i.e. complex software systems build around a persistent database. In contrast to pure database conceptual modeling, the modeling of database applications has to put emphasis on the *dynamics* of the application software. In contrast to classical software specification, we have to offer powerful techniques to desribe large complex–structured databases and to integrate the handling of persistent objects.

In the next section, we will discuss the special requirements arising from the aim to describe database applications in an appropriate fashion. The third section presents a layered approach for the conceptual modeling of database applications. Both language and presentation issues as well as the formal semantics are briefly discussed. In the fourth section, we discuss methodology issues like modularization of conceptual models using an object–oriented approach or tools and methods desirable to support the conceptual modeling process.

It is impossible to give a complete conceptual design language and / or methodology in this paper. However, we think that several established design methodologies can be integrated in the design of database applications. The following sections should therefore be regarded as a survey how to structure the conceptual modeling of a database application in some design framework and as a guide to choose appropriate description formalisms if needed. The interested reader should consider the cited references for more details on related approaches.

2 Modeling of Database Applications

In this section, we will sketch some special requirements for conceptual modeling of database applications. A *database application* is a software system build around one or more persistent databases. In general, the database application may integrate further software components like CAD software and may interact with several users and / or other applications.

The modeling of a database application has to integrate two main modeling tasks, the modeling of the *database structure* and the design of the *application dynamics*. Both tasks are strongly connected and influence each other.

The *structural modeling* can be seen as the classical database design process describing possible collections of persistent objects (the database) in an appropriate formalism. There are several conceptual database design methodologies proposed in the literature, which are discussed in section 3.2. However, we have some special requirements arising from the integration with the application dynamics specification.

First of all, we have to integrate the specification of *abstract data types* into the structural modeling. Established database modeling frameworks like the ER model usually concentrate on the stored objects only. In application software on the other hand, we have the richness of arbitrary user–defined values and operations on these values as known from modern programming languages. These data types must be semantically integrated into the database specification framework. A specification framework for abstract data types is discussed in section 3.1. In application software specifications we prefer strongly typed semantic domains which should be considered in the semantics definition of the used conceptual data model. Furthermore, the design framework should support arbitrary integrity constraints and rules for deriving implicitly stored information.

The *database application dynamics* consists of two components, of the database dynamics describing the correct database evolutions and of the application–specific processes. The *database evolution* can be described completely without referring to specific application activities, for example using a temporal logic framework. This approach is discussed in section 3.3. An alternative description of the allowed evolutions specifies the allowed update *actions* in a high–level framework as presented in section 3.4. Both evolution descriptions together describe the correct evolutions of the specified database through the course of time.

As last design component, the *application–specific processes* have to be specified in an appropriate framework. This task can be accomplished using a framework taken from software engineering methods for process systems. The database itself must be modeled as a special persistent process in a way compatible to the database semantics chosen for database structure and evolution specification. Examples for application–specific processes are controlled user interaction, multi–database applications and long–term transactions known to be relevant in the engineering database area.

3 Layered Conceptual Models

As motivated in the previous sections, we divide a conceptual model of a database application into several *specification layers* following [SFNC84, CaS87, HNSE87]. Those layers use specification formalisms building a hierarchy of semantic interpretation structures using a hierarchy of specification languages. At the level of describing database evolutions, the strict hierarchy of layers is broken because we have two independent description formalisms building complementary specifications of the same target. Concepts of lower layers are integrated appropriately into upper layers. The hierarchy of layers is shown in figure 1.

We will present the basic ideas of such a layered approach in the following sections. A more detailed presentation of the four lower layers (which mainly desribe the database part of an application) can be found in [EGH+90].

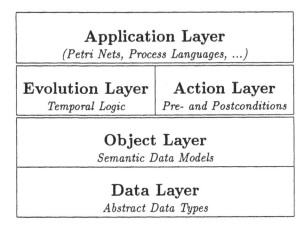

Figure 1: Layers of Conceptual Model Descriptions

3.1 Data Layer

On the *data layer* we have to describe the state–independent basic data structures. In software specifications, these data structures are encapsulated in *abstract data types* together with related functions and predicates on the data elements. These state–independent data structures define the structures of basic data items stored in the database as properties of persistent objects. They are also known as *printable* or *lexical* object types in the database modeling literature (see [EGH+90] for a more detailed discussion and a literature overview).

Usually, implemented database management systems nowadays support only a small set of standard data types — in contrast to the area of programming, where user–defined data structures are supported by most modern programming languages. A specification formalism merging the fields of these both classical disciplines has to offer more powerful specification concepts as done in classical database design. In recent years, the development of *extensible* database management systems tries to bridge this gap at the level of database implementations, too.

Examples for user–defined data types are geometric data types like `point`, `line` or `circle` with related operations like `circle_cut` or `distance`. Other examples are enumeration types, data types for engineering applications like `vector` as well as types for large unstructured data like bitmap pictures or video sequences.

As a specification formalism for abstract data types we can choose one of the well–established description formalisms which can be found in the related software engineering literature. Specification of abstract data types is not specific to database applications, and we will not go into detail here.

An established specification formalism is the *algebraic* specification of abstract data types using equational specification (see [EhM85] for a textbook). Other alternatives are the use of recursive function definitions or directly defining the functions in an imperative language. The following example shows part of the specification of the geometric data type `point` done in the equational framework.

Example 3.1 The geometric type point together with related operations can be specified explicitly as follows :

```
DATATYPE point BASED ON real;
SORTS point;
OPERATIONS distance :        (point × point):  real;
           xcoord, ycoord :  (point):  real;
           createpoint :     (real × real):  point;
           add :             (point × point):  point;
           ...
VARIABLES p,q :  point;
          x,y,x1,y1 :  real;
EQUATIONS
   x = xcoord(createpoint(x,y));
   y = ycoord(createpoint(x,y));
   distance(createpoint(x,y),createpoint(x1,y1))
     = sqrt((x-x1)*(x-x1) + (y-y1)*(y-y1));
   add(p,q)
     = createpoint(xcoord(p)+xcoord(q),ycoord(p)+ycoord(q));
   ...
```

$\square$

The formal semantics of such a specification is given by heterogeneous algebras, i.e. by a set of values for each sort symbol and a function on these values for each function symbol.

To support the usual mechanisms to construct new types from already defined ones, we additionally have a collection of *parameterized data type constructors* like set or list construction. With each of these constructors a family of operations is associated. For example, the operations in, insert and union are associated with the set constructor (among others). These constructors can be used to build a family of polymorphic types to simplify the use of data types in specifications. The data type constructors are also used for defining the result structures of queries and for the definition of the type of multi–valued attributes, too.

3.2 Object Layer

At the *object layer*, the *consistent database states* are described. A database state can be seen as a snapshot of the persistent objects representing the information stored in the database. The modeling of the object layer is closely related to the classical design techniques for database structure using conceptual data models like the ER model, semantic data models or object–oriented models. In fact, a modern conceptual data model proposal should try to combine the pros of those succesful approaches. The conceptual description of database object collections has been a main field of database research for several years. Therefore, we will not give an elaborated literature overview but show some entry points into the related literature only.

The description of the object layer consists of two parts, the description of the proper database *structure* in terms of a data model and the description of *correct extensions* of this structure definition in terms of integrity constraints.

As mentioned before, we want to describe *collections of persistent objects carrying information*. The information carried by objects is expressed in terms of data–valued object properties (called *attributes*) and relationships between objects. These concepts are the basic modeling concepts offered by the Entity–Relationship Approach [Che76]. Experiences with modeling complex applications, especially in the area of so–called 'non-standard applications' like engineering databases [BaK85, KaL88], have shown that we need further concepts to support special relationships between persistent objects like ISA or PART_OF relations. These additional concepts originating from the development of so–called *semantic data models* [HaM81, HuK87] can be integrated into the ER approach [EWH85, HNSE87, ElN89, EGH+90]. Recently, the discussion of *object models* [DaD86, ACM87, Dit88, ABD+89, Bee89, KiL89, ZdM89] has brought new aspects to the discussion on appropriate description of conceptual data models, among them inheritance along subclass hierarchies and temporal object identity independent of current attribute values [KhC86, ABD+89]. Another interesting extension is to use rules to derive implicitly expressed objects, properties and relationships.

We think that all the benefits of these modeling approaches should be supported by a specification language for the object layer — as long as we can give a formal semantics to the introduced concepts. For example, at the moment we would restrict inheritance to subclass hierarchies allowing a direct inheritance definition without introducing compli-cated rules for problematic cases. The formal semantics of a conceptual database schema is given by the set of all proper database states being extensions of this schema. Such database states are interpretation structures of a many–sorted predicate logic containing the semantics of the data layer as a subalgebra.

It should be mentioned here, that each schema of a conceptual data model defines the *signature* of a many–sorted predicate logic, where the sort symbols are given by the object sorts (and data type sorts, too) and functions and predicates are induced by the attribute and relationship definitions of the schema [HoG88]. This logic is the basis for query formalisms and results in a language for integrity constraints. Another language induced by the schema is a language for elementary updates [EGH+90].

After these considerations on the object layer as a whole, we will now look a little closer to the single modeling concepts offered by a suitable specification language. We will give a short list of the basic modeling concepts we think being mandatory for specifying the object layer appropriately :

- The first modeling primitive is the concept of abstract entities called *objects* or *entities*. Objects are abstract in the sense that they can only be observed by the values of their properties. Properties are data– or object–valued functions for objects and are called *attributes*.

- Objects with the same set of properties can be grouped into *object types*. Examples for object types are the types PERSON or COMPANY with corresponding data–valued attributes, for example, name of type string or location of type point. An example for an object–valued attribute would be the attribute manager of type PERSON associated with an object type DEPARTMENT. Object–valued attributes can often be adequately modeled by functional relationships or complex object construction, too.

- Objects are abstract entities observable by their attributes only. To distinguish different objects having the same properties, we have to introduce an *object identification mechanism* [KhC86]. Object identity can be specified explicitly by *key functions*, i.e. by choosing some object properties as object 'separators' inside one object type [EDG86, HNSE87, SSE87]. An alternative solution is to introduce an *implicit object identity* as a property of the data model as it is done in some object–oriented approaches [ABD+89, Bee89, KiL89].

- With an object type we associate the class of currently existing objects of this type. Usually, these classes are disjoint. But there are several interesting cases where this intutively is not the case. In these cases we talk about *type* or *class construction* by generalization, specialization or partition. Constructed classes inherit the identification from their base types. For the formal semantics of type constructions see [AbH87, HNSE87].

  - *Specialization* is used to build a subclass hierarchy, for example starting with the type PERSON and defining MANAGER and PATIENT as independent subclasses of PERSON. Specialization induces a subset relation (ISA hierarchy) between the current object class populations and a inheritance of properties of the input type.
  - *Partition* is a special case of specialization where a type is partitioned into several disjoint subclasses. An example is the partition of PERSON into WOMAN and MAN.
  - *Generalization* works the other way round — several input classes are generalized into a new class. An example is the generalization of PERSON and COMPANY into LEGAL_PERSON.

- Another modeling concept known from the ER approach are arbitrary *relationships* between objects, for example the relationship Works_For between persons and companies. There are several interesting special relations between objects, which should be explicitly modeled in a specification. Examples are the already mentioned ISA relation or functional relationships (being equivalent to object–valued attributes in the binary case).

- Another special relation which should be made explicit is the PART_OF relation leading to the notion of *complex objects*. Especially in engineering applications, the appropriate definition of complex objects is a mandatory feature of a conceptual data model [BaK85, DGL86, KaL88]. There are several properties associated with the notion of complex objects, among them weak object types (a subobject can not exist outside its superobject), the distinction between disjoint and non-disjoint complex objects and the problem of update propagation for complex objects.

Until now, we have concentrated on the proper structure of our object collections. If we want to express additional restrictions and knowledge from the application area in the object layer specification, we have to state *integrity constraints* restricting the correct states. Some common integrity constraint patterns are usually directly supported by specific language features, for example cardinality constraints on relationships. On the conceptual level, other constraints are formulated in a first–order logic induced by the conceptual schema.

Example 3.2 The constraint that each employee of a department has to earn less than the manager of her / his department can be formulated as follows :

```
FOR ALL (P : PERSON), (M : MANAGER), (D : DEPARTMENT) :
   ( Works_For(P,D) AND D.manager = M AND NOT P = PERSON(M) )
      IMPLIES P.salary < M.salary;
```

In this example, we have used an explicit conversion of a MANAGER object into a PERSON object along the subtype hierarchy defined by a specialization relationship. □

Another way to express additional application area knowledge is the use of *rules* to derive information from explicitly stored objects. For the modeling of database states, we use a model–based semantics because we think it is appropriate for specifying database states being implemented by concrete interpretations of a data model. Therefore rules are used only in a restricted way, namely to compute derived attributes, objects and relationships in a determined fashion. A commonly used derivation is the definition of so–called *computed* or *derived attributes* by a data–valued function. We present the definition of a derived relationship as a more general example.

Example 3.3 As an example we look at a rule deriving a relation expressing the transitive closure of the management relation. The recursive definition starts with an object–valued attribute **manager** of PERSONs storing the direct manager of a person.

```
DERIVED RELATIONSHIP Manages(MANAGER,PERSON)
IS DEFINED BY
  VARIABLES (P : PERSON), (M,M': MANAGER) :
      Manages(M,P) IF P.manager = M;
      Manages(M,P) IF Manages(M,PERSON(M')) AND Manages(M',P);
```

The rule is formulated in a syntax compatible to the constraint language. The IF keyword denotes an implication from right to left, and variables are universally quantified.
□

There is a close relationship between rules and integrity constraints. If derived information is modeled explicitly on the object layer, the derivation rules are read as special integrity constraints. On the conceptual level, both views are equivalent and need not to be distinguished. However, the modeling of derivation rules is an important part of the application modeling and should be supported by appropriate language constructs.

3.3   Evolution Layer

Until now, we have described the static aspects of database states only. The next specification layer, the *evolution layer*, specifies the *temporal evolution* of the persistent objects. This is done completely without referring to the concrete modification actions changing the stored information. The reference time scale is the causal time induced by the sequence of database modifications.

The semantics domain to be specified is the set of *correct database state sequences*. This is done independently from concrete transactions or application processes. The temporal evolution of the stored information is specified by restricting the *life cycles* of persistent database objects. Such restrictions are called *dynamic* or *temporal* constraints. With other words, we state which long–term evolutions of object (or object combinations) properties and relations are desired. Examples of such long–term dynamic constraints are

- *"Salaries of employees must not decrease."*

- *"Airplanes have to be maintained at least once in a year, but at the latest every 50.000 miles."*

- *"Employees have to spend their yearly holidays until May of the following year."*

There are several specification formalisms for such dynamic constraints proposed in the literature :

- *Temporal logic* specifications offer a descriptive formalism for temporal constraints. Their semantics is directly expressed using sequences of predicate logic interpretations, i.e., of database state sequences.

  Several temporal logic dialects for temporal constraints are proposed in the literature, for example in [Ser80, CCF82, ELG84, CaS87, Saa88, Lip89, Saa90].

- An alternative, more procedural way to express temporal constraints is to use transition automata or simple Petri nets. This technique is, for example, proposed by [EKTW86].

Both approaches are equivalent in the sense that a given specification using one approach can be automatically compiled into a specification using the alternative approach (see [LiS87, Saa88] for the transformation from temporal logic into automata). This transformation into transition automata can be interpreted also as a transformation into *transitional constraints* restricting local state transitions instead of whole state sequences. As an interesting extension of dynamic constraints, [WMW89] additionally proposes to distinguish between dynamic constraints and *deontic constraints* separating the correct database sequences and the desired temporal evolutions.

It should be noted, that both approaches need a formal semantics of *temporal object identity* because temporal logic formulae or transition automata are formulated locally for single objects changing their properties during database evolution. For example, a PERSON object remains the same object even if all its observable properties are changing (assuming an implicitly given temporal object identity as offered by object–oriented models [KhC86]). We give two examples for temporal logic constraints.

Example 3.4 The dynamic constraint that salaries of employees must not decrease is formulated as follows :

```
FOR ALL (E : EMPLOYEE) (s :  integer):
  ALWAYS (E.salary = s IMPLIES ALWAYS NOT E.salary < s );
```

The temporal operator ALWAYS denotes a temporal quantification over *all* future states. The first ALWAYS defines the bound subformula as an invariant, i.e., the formula must be satisfied for an inserted PERSON object in all future database tail sequences. The inner implication states that if once the salary of an EMPLOYEE is equal to an integer value s, it must be greater or equal to s for all future states (due to the inner quantification by ALWAYS). □

Example 3.5 The second dynamic constraint states that salaries of employees must not decrease while working at the same company — even if she / he has worked for another company in the meanwhile.

```
FOR ALL (E : EMPLOYEE) (s :  integer) (C : COMPANY):
  ALWAYS ((E.salary = s AND Works_For(E,C))
      IMPLIES ALWAYS ( Works_For(E,C) IMPLIES NOT E.salary < s) );
```

The interesting point of the second example is that this constraint implicitly uses historical information, namely the former salaries of persons earned at companies, even if the explicit information that a specific person had worked for a company in the history is not modelled in the object schema directly. The identification and consideration of such additional object structure induced by dynamic constraints is an important part of the conceptual database design process. This problem is discussed in more detail in [HüS89, HüS90]. □

3.4 Action Layer

In the previous subsection, we have presented a specification method to describe database evolutions independently of concrete modification transactions. The *action layer* offers the complementary description of database sequences in terms of correct *database state transitions* by so–called *actions*.

Actions are schema–specific database updates, i.e. functions from database states into new correct database states. They are the elementary building blocks of transactions preserving integrity.

Examples of actions are insertion of an employee, or a salary upgrade while respecting the constraints on employee's salaries, or a flight reservation in a travel agency database.

There are several proposals for specification techniques for database actions. Since an action is a function on database states, we can use specification mechanisms for functions on values of a complex structured data type, for example algebraic specification as proposed by [FuN86]. However, this approach neglects somehow our more abstract view on database states as interpretation structures of a logic theory. We prefer to use specification formalisms interpreting action specifications as a relation between first order logic models fitting to the semantic domains used for the evolution layer.

A natural way to describe transitions between interpretation structures is to use *pre– and postconditions*. This decriptive style of action specifications fits well to the use of temporal logic for describing database evolutions [Lip89, SaL89, Saa90]. A detailed language proposal independent of a fixed data model and its formal semantics can be found in [Lip88, Lip89]. A language proposal for an extended ER model is presented in [EGH+90].

Pre– and postconditions are a restricted form of a modal or action logic using explicit logic operators referring to actions. Such specification frameworks are used in [KMS85, FiS88] to specify actions using arbitrary modal / action logic formulae.

An example of an action specification using pre– and postconditions is the action **FireEmployee** specified in the following example.

Example 3.6 The action specification **FireEmployee** removes a person from the database if she / he is not currently manager of another person.

```
ACTION FireEmployee (person_name :  string);
  VARIABLES P : PERSON;
  PRECONDITION P.name = person_name IMPLIES
              NOT EXISTS (PP : PERSON) P = PERSON(PP.manager);
  POSTCONDITION NOT EXISTS (P : PERSON) P.name = person_name;
```

The object variable P is implicitly universally quantified over all currently existing persons. □

A specification using pre- and postconditions describes the desired effects of an action only. There are usually several transition functions between database states satisfying such a specification. To capture desired and undesired side effects of state transitions satisfying the specification, we need two implicit rules to choose *minimal correct transitions* as a standard semantics.

- The *frame rule* states that an action effect should be as minimal as possible. The existence of a minimal transition is, however, an undecidable problem, for example if we have disjunctive postconditions. An elaborate discussion of the frame rule and related problems can be found in [Lip88, Lip89]. The frame rule forbids undesired side effects of actions ('no junk').

- The *consistency rule* states that each action has to obey the (static and dynamic) integrity constraints. It handles the desired side effects of actions like update propagation.

Both rules work complementary : The consistency rule extends the action specification e.g. by additional postconditions to guarantee integrity. The frame rule on the other hand has to add invariants guaranteeing that only object modifications can occur which are explicitly enforced by postconditions or by the need of integrity preservation.

3.5 Process Layer

In [Lip88, EGH⁺90], the presented four lower specification layers are identified as being relevant to describe the structural and dynamic properties of a database as a stand–alone component. However, to describe *database applications* as a software system consisting of a (central ?) database and further components, we have to add an additional layer describing these system components and their interaction in a suitable framework. Moreover, this description framework should be compatible to the semantics of the pure database description layers.

At the *process layer*, we describe a database application as a *collection of interacting processes*. The database described using the four lower layers is handled as one special persistent process where the actions determine the event alphabet of the process. The database process is purely reactive, i.e. actions are triggered from other processes only. This approach is powerful enough to handle distributed applications, formal user modeling and multiple database applications in the same framework.

Semantically, the database process can be described as a linear life cycle over the event alphabet together with an observation function mapping prefixes of the life cycle into database states. This semantics is conform with the semantic models used for the pure database specification, i.e. with linear sequences of database states.

The database is only one process among others which together build the database application. Conceptually, we regard the application as consisting of several independent software components communicating by sending and receiving messages. Examples for such components are

- *interaction interfaces* communicating with application users using an application–specific communication protocol, for example based on formulars or windows,

- long–term *engineering transactions* performing complex activities in cooperation with several users and databases,

- other integrated *software systems* like CAD systems,

- and last, but not least, several *data and object bases* possibly implemented using different DBMSs and data models.

The formal specification of interacting processes is still a vivid field of software engineering research. We propose to adopt one of the various approaches to describe such systems being around in the literature, for example process definition languages [Hoa85] or net–based formalisms [RiD82, Rei85, Lau88].

4 Design Issues

In the previous section, we presented a multi–layered framework for conceptual modeling. A conceptual model is specified using different specification formalisms on each layer. The semantics of a layered specification is formalized using different techniques like mathematical algebras, first order theories, sequences of first order logic interpretation structures and communicating processes integrated in a layered construction of semantic interpretation structures. This approach handles the problem of structuring the specification framework, but, however, we still have to solve the problems arising with the complexity of specification documents.

4.1 Complexity and Consistency of Conceptual Models

There are two main problems related with the complexity of specification documents. Firstly, there is the problem of handling very large design documents during the design process. It is well known in the area of database design, that the structure of the object level alone can easily reach a complexity making it impossible to handle the design documents without an additional modularization — flat ER diagrams with hundreds of entities and relationships cannot be appropriately handled during system design. These problems are multiplicated in a framework where several static constraints, temporal constraints and action specifications are connected with each structural object of the database. We need both a structuring mechanism breaking the complete specification into handy pieces as well as computer aided support for large design documents. These problems are discussed in subsections 4.2 and 4.3.

A further point where we need formal support is *consistency checking* for application specifications. The single layers use different specification logics describing different semantic interpretation structures which are in the end combined into interpretation structures for the complete database application.

Of course, the different layer specifications can contradict each other making the complete specification unsatisfiable. We just enumerate some of the critical layer combinations which should be tested on compatibility.

1. Firstly, we should not forget that each layer itself can be inconsistent — a problem which is really hard to handle even for the lowest structural specification layers. An example is the satisfiability problem for a set of first order logic constraints.

2. A critical point is the consistency of static and dynamic integrity constraints together. This can be tested up to a certain degree with methods presented in [Saa88, SaL89, Saa90]. The basic idea is to interpret static constraints as special temporal invariants, then construct a transition automaton out of the combined constraints and test the resulting automaton if it accepts at least one state sequence.

3. An interesting combination is the interplay between the evolution layer and the action layer. There are several sources of inconsistency in the combination of both layers:

   • Even if both layers are satisfiable, i.e. if there exists at least one state sequence for each layer satisfying the specifications, the combined layers may be unsatisfiable. An example is the situation where a satisfiable temporal requirement cannot be fulfilled by any action, because no action does perform the update necessary for constraint satisfaction.

   • A specific action specification can be never executable because its postcondition always violates integrity constraints.

   Some ideas on checking combined specifications for detecting such situations are presented in [Saa90]. An alternative way is to integrate both static and temporal constraints into action specifications as proposed by [Lip88, LiS88, Lip89] and then check the extended action specifications on satisfiability.

4. Finally, even if the database described at layers one to four is consistent, the application processes may prevent the satisfaction of temporal constraints forever, or never deliver the precondition for an action, or run into a deadlock because the system reached a state where no action can be executed without violating integrity.

It should be clear even after these few short remarks that global consistency tests for large specification documents usually cost a lot.

4.2   Object–Oriented Structuring

Recently, the *object–oriented paradigm* is transfered to conceptual modeling of information systems and database applications, too. First examples of object–oriented specification languages and methods can be found in [SSE87, SFSE89] and in [Wie90]. Main motivations for introducing object–oriented techniques into conceptual modeling are the possibilities to modularize conceptual models and to encapsulate structure and behaviour in objects.

To distuinguish the notion of an object encapsulating structure and behaviour from our previous purely structural object notation of section 3.2., we will sometimes use the terminology *object unit* if we denote a structural object together with its behaviour as introduced as the basic design unit in [SFSE89].

The basic principles of object–oriented conceptual modeling can be summarized as follows:

> An *object* has a current state expressed by *attribute* values which can be modified by object–specific *events* only. It can be created by a birth event and removed from the object base by a death event. The semantic interpretation of an object specification is an *observed process*, i.e. a collection of correct event life cycles together with an observation function mapping prefixes of life cycles into attribute values [SSE87, SeS89, ESS90]. Object specifications consist of a set of attribute constraints (static and temporal), of the definition of the attribute evaluation for each event and of a process specification in a suitable framework.
>
> Objects *interact* with each other by event sharing or event calling. An *object society* is a collection of interacting objects. A simple *object type* consists of a surrogate space (i.e. a data type delivering possible surrogate values) together with a template describing the schema of the objects in the type extension. There are several modeling primitives to construct complex objects, specializations and generalizations etc. as known from semantic data models similar to the constructs discussed in section 3.2.

A more elaborated introduction into object–oriented conceptual modeling can be found in [SFSE89].

How does this approach fit into the framework for conceptual modeling presented in chapter 3 ? The *semantics* of an object unit is an observed process, i.e. a linear process together with observable attributes. These observable attributes can be complexly structured, for example for complex aggregated objects. *Therefore, the semantics domain for such object units is the same as for the semantics of complete database processes as discussed in section 3.5.* A four–layer description of a database can be interpreted as a single object unit, and the application processes can be seen as other object units communicating with each other.

This insight in the relation between layered conceptual modeling and object–oriented design allows us to use object–oriented design techniques in coexistence and cooperation with established design techniques. We can split a database into smaller database units on the conceptual level semantically interpreted as interacting object units. Such database units can be designed and analysed independently and later on aggregated to the complete database in a semantics preserving way [SFSE89, FSMS90]. Additionally, the specifications of these units can be formally refined using techniques known from formal implementation of abstract data types [EhS90].

This splitting into smaller units can be repeated until each structural object of the object layer together with related constraints and actions is encapsulated into one object (type) unit. Object interaction arises from actions and constraints involving more than one object unit at a time. All modeling constructs presented at the object layer can be extended to encapsulated object units [SFSE89].

Summarizing, object–oriented design techniques allow to structure a conceptual database schema into smaller modules later on aggregated to the complete database schema. The global database is the aggregation of all object modules. We gain the following benefits from these structuring principles :

- We have a structuring principle for conceptual models with a well–defined formal semantics [SSE87, SeS89, EhS90, FSMS90].

- The object units make local consistency checking possible [FSMS90].

- The semantics of object units fits into the semantics approach chosen for the application layer.

4.3 Tools & Methods

Even for medium–sized applications, a formal support of the conceptual design by appropriate methods and tools is necessary. We will not go into details of database design environments but give only an enumeration of desired features of such systems supporting conceptual database modeling. These desired features are derived from the description of a database design environment currently under development [EHH+89].

An integrated environment should offer at least the following tools to support conceptual design of database applications :

- A family of *syntax oriented editors* for creating and manipulating specification documents. If possible, these editors should offer a graphical presentation of the document, for example an extended ER diagram for the object layer or a visualization of the corresponding transition automata for temporal logic formulae. Both navigation inside one layer and inter–layer navigation (coming from object types to actions manipulating them) must be supported.

- Several tools for *consistency checking* as mentioned already in section 4.1. As far as possible, these tools should be integrated into the corresponding editor facilities.

- A *prototyping component* making an early validation of the conceptual model possible. Crucial points are the prototyping of data type and action specifications and finally the animation of the process layer.

All these components should be integrated into one environment having a uniform user interface for the single components. Of main importance is the support of incremental design allowing consistency checking and prototyping already for partially specified applications.

5 Conclusions and Future Work

We have presented a multi–layered approach to conceptual design of database applications following the proposals of [SFNC84, CaS87, HNSE87] for database models. The concepts of values, persistent objects, temporal object evolution, update operations and arbitrary application processes are separated clearly into specification layers integrated into one uniform semantical framework.

The design formalisms for the lower four layers, i.e. for the pure database component, has evolved to a stabilized design framework [EGH+90]. For the application layer, however, we still have to develop design techniques fitting well to the classical database design process. The different existing formalisms and methods for process design must be inspected carefully for their suitability for designing data–intensive applications, an aspect of process design which has played no main role in process design until now.

The major challenge for the near future is the integration of object–oriented design techniques and established database design methodology as indicated in section 4.2. The advantages of modularization and encapsulation of structure and behaviour into object units must be transfered to conceptual database modeling and transaction specification techniques. The consequences for design methodologies and supporting tools have to be explored, too.

Acknowledgements

The ideas presented in this paper are mainly influenced by discussions and research projects performed in the last years at the Technical University of Braunschweig, F.R.G. I wish to thank my colleagues involved in this work, namely Hans–Dieter Ehrich, Gregor Engels, Martin Gogolla, Uwe Hohenstein, Klaus Hülsmann, Udo Lipeck, Perdita Löhr–Richter and Leonore Neugebauer. The reflections on object–oriented modeling presented in section 4.2 are mainly influenced by the cooperation in the ESPRIT BRA WG 3023 (ISCORE) and the discussions with other project partners, namely José Fiadeiro, Ralf Jungclaus, Amilcar Sernadas and Cristina Sernadas. Thanks to Ralf Jungclaus for his helpful comments on an early version of this paper.

References

[ABD+89] Atkinson, M., Bancilhon, F., DeWitt, D., Dittrich, K., Maier, D., Zdonik, S.: The Object–Oriented Database System Manifesto. *Proc. First Int. Conf. on Deductive and Object-oriented Databases DOOD'89 (W.Kim, J.-M.Nicolas, S.Nishio, eds.).* Kyoto, 1989, **40–57**.

[AbH87] Abiteboul, S., Hull, R.: IFO — A Formal Semantic Database Model. *ACM Transactions on Database Systems.* Vol. 12, No. 4, 1987, **525–565**.

[ACM87] *ACM Transactions on Office Information Systems.* Special Issue on Object–Oriented Systems. Vol. 5, No. 1, 1987.

[BaK85] Batory, D.S, Kim, W.: Modeling Concepts for VLSI CAD Objects. *ACM Transactions on Database Systems.* 10 (3), 1985, **322–346**.

[Bee89]   Beeri, C.: Formal Models for Object Oriented Databases. *Proc. First Int. Conf. on Deductive and Object-oriented Databases DOOD'89 (W.Kim, J.-M.Nicolas, S.Nishio, eds.).* Kyoto, 1989, **370–395**.

[BMS84]   Brodie, M.L., Mylopoulos, J., Schmidt, J.W.: On Conceptual Modelling — Perspectives from Artificial Intelligence, Databases, and Programming Languages. Springer 1984.

[CaS87]   Carmo, J., Sernadas, A.: A Temporal Logic Framework for a Layered Approach to Systems Specification and Verification. *Proc. IFIP WG 8.1 Conf. on "Temporal Aspects in Information Systems".* Sophia–Antipolis 1987, **31–46**.

[CCF82]   de Castilho, J.M.V., Casanova, M.A., Furtado, A.L.: A Temporal Framework for Database Specification. *Proc. Int. Conf. on Very Large Databases.* Mexico City 1982, **280–291**.

[Che76]   Chen, P.P.: The Entity–Relationship Model — Towards a Unified View of Data. *ACM Transactions on Database Systems.* Vol. 1, No. 1, 1976, **9–36**.

[DaD86]   Dayal, U., Dittrich, K.(eds.): Proc. Int. Workshop on Object–Oriented Database Systems. IEEE Computer Society, Los Angeles 1986.

[DGL86]   Dittrich, K., Gotthard, W., Lockemann, P.C.: Complex Entities for Engineering Applications. *Proc. 5th Int. Conf. on Entity Relationship Approach.* North–Holland, Amsterdam 1986.

[Dit88]   Dittrich, K.(ed.): Advances in Object–Oriented Databases. LNCS 334, Springer–Verlag, Berlin 1988.

[EDG86]   Ehrich, H.-D., Drosten, K., Gogolla, M.: Towards an Algebraic Semantics for Database Specification. *Proc. 2nd IFIP Work. Conf. on Database Semantics "Data and Knowledge" (DS-2), Albufeira 1986 (R.A.Meersmann, A.Sernadas, eds.).* North–Holland, Amsterdam 1988, **119–135**.

[EGH+90]   Engels, G., Gogolla, M., Hohenstein, U., Hülsmann, K., Löhr–Richter, P., Saake, G., Ehrich, H.-D.: Conceptual Modelling of Database Applications Using an Extended Entity–Relationship Model. To appear as: Informatik–Bericht, Techn. Univ. Braunschweig, 1990.

[EHH+89]   Engels, G., Hohenstein, U., Hülsmann, K., Löhr–Richter, P., Ehrich, H.-D.: CADDY: Computer Aided Design of Non–Standard Databases. To appear in: *Proc. 1st Int. Conf. on System Development Environments and Factories.* Berlin 1989.

[EhM85]   Ehrig, H., Mahr, B.: Fundamentals of Algebraic Specification 1. Equations and Initial Semantics. Springer–Verlag, Berlin 1985.

[EhS90]   Ehrich, H.-D., Sernadas, A.: Algebraic View of Implementing Objects over Objects. *Proc. REX Workshop Stepwise Refinements of Distributed Systems : Models, Formalisms, Correctness (W.deRoever, ed.).* Springer–Verlag 1990.

[EKTW86]   Eder, J., Kappel, G., Tjoa, A.M., Wagner, R.R.: BIER : The Behaviour Integrated Entity Relationship Approach. *Proc. of the 5th Int. Conf. on Entity–Relationship Approach (S.Spaccapietra, ed.).* Dijon 1986, **147–166**.

[ELG84]   Ehrich, H.-D., Lipeck, U.W., Gogolla, M.: Specification, Semantics and Enforcement of Dynamic Database Constraints. *Proc. Int. Conf. on Very Large Databases.* Singapore 1984, **301–308**.

[ElN89]   Elmasri, R., Navathe, S.B.: Fundamentals of Database Systems. Benjamin / Cummings Publ., Redwood City 1989.

[ESS90]   Ehrich, H.-D., Sernadas, A., Sernadas, C.: From Data Types to Object Types. *J. Inf. Process. Cybern. EIK*. Vol. 26, No. 1/2, 1990, **33–48**.

[EWH85]   Elmasri, R.A., Weeldreyer, J., Hevner, A.: The Category Concept : An Extension to the Entity–Relationship Model. *Data & Knowledge Engineering*. Vol. 1, 1985, **75–116**.

[FiS88]   Fiadeiro, J., Sernadas, A.: Specification and Verification of Database Dynamics. *Acta Informatica*. Vol. 25, Fasc. 6, 1988, **625–661**.

[FSMS90]   Fiadeiro, J., Sernadas, C., Maibaum, T., Saake, G.: Proof–Theoretic Semantics of Object Oriented Specification Constructs. To appear in: *Proc. IFIP TC2 Work. Conf. on Database Semantics : Object Oriented Databases (DS-4)*. Windermere (UK), July 1990.

[FuN86]   Furtado, A.L., Neuhold, E.J.: Formal Techniques for Data Base Design. Springer–Verlag, Berlin 1986.

[HaM81]   Hammer, M., McLeod, D.: Database Description with SDM : A Semantic Data Model. *ACM Transactions on Database Systems*. Vol. 6, No. 3, 1981, **351–386**.

[HNSE87]   Hohenstein, U., Neugebauer, L., Saake, G., Ehrich, H.-D.: Three Level Specification of Databases Using an Extended Entity Relationship Model. *GI/ÖGI/SI-Fachtagung "Entwurf von Informationssystemen, Informationsbedarfsermittlung und -analyse", Linz (R.R.Wagner, R.Traunmüller, H.C.Mayr, eds.)*. Informatik–Fachbericht Bd. 143, Springer–Verlag, Berlin 1987, **58–88**.

[Hoa85]   Hoare, C.A.R.: Communicating Sequential Processes. Prentice–Hall, Englewood Cliffs 1985.

[HoG88]   Hohenstein, U., Gogolla, M.: Towards a Semantic View of an Extended Entity-Relationship Model. Informatik–Bericht Nr. 88–02, Techn. Univ. Braunschweig 1988.

[HuK87]   Hull, R., King, R.: Semantic Database Modelling : Survey, Applications, and Research Issues. *ACM Computing Surveys*. Vol. 19, No. 3, 1987, **201–260**.

[HüS89]   Hülsmann, K., Saake, G.: Theoretical Foundations of Handling Large Substitution Sets in Temporal Integrity Monitoring. Informatik–Bericht Nr. 89-04, Techn. Universität Braunschweig 1989.

[HüS90]   Hülsmann, K., Saake, G.: Representation of the Historical Information Necessary for Temporal Integrity Monitoring. *Proc. Int. Conf. on Extending Database Technology EDBT'90 (F.Bancilhon, C.Thanos, D.Tsichritzis, eds.)*. LNCS 416, Springer–Verlag, Berlin 1990, **378–392**.

[KaL88]   Karl, S., Lockemann, P.C.: Design of Engineering Databases : a Case for More Varied Semantic Modelling Concepts. *Information Systems*. Vol. 13, No. 4, 1988, **335–358**.

[KhC86]   Khoshafian, S., Copeland, G.P.: Object Identity. *ACM Proc. Conf. on Object-Oriented Programming Systems, Languages, and Applications*. Portland, Sept.86. Reprint in: [ZdM89], **37–46**.

[KiL89]   Kim, W., Lochovsky, F.H.(eds.): Object–Oriented Concepts, Databases, and Applications. Addison–Wesley Publ., Reading 1989.

[KMS85]   Khosla, S., Maibaum, T.S.E., Sadler, M.: Database Specification. *Proc. IFIP Conf. on Datbase Semantics (DS–1), 1985 (T.B.Steel, R.Meersmann, eds.).* Noth Holland, Amsterdam 1986, **141–158**.

[Lau88]   Lausen, G.: Modeling and Analysis of the Behavior of Information Systems. *IEEE Transactions on Software Engineering.* Vo. 14, No. 11, 1988, **1610–1620**.

[Lip88]   Lipeck, U.W.: Transformation of Dynamic Integrity Constraints into Transaction Specifications. *Proc. 2nd Int. Conf. on Database Theory 1988 (M.Gyssen et al, eds.).* LNCS 326, Springer–Verlag, Berlin 1988. **322–337**.

[Lip89]   Lipeck, U.W.: Zur dynamischen Integrität von Datenbanken: Grundlagen der Spezifikation und Überwachung (Dynamic Integrity of Databases: Fundamentals of Specification and Monitoring). Informatik–Fachbericht IFB 209, Springer–Verlag, Berlin 1989.

[LiS87]   Lipeck, U.W., Saake, G.: Monitoring Dynamic Integrity Constraints Based on Temporal Logic. *Information Systems.* Vol. 12, No. 3, 1987, **255–269**.

[LiS88]   Lipeck, U.W., Saake, G.: Entwurf von Systemverhalten durch Spezifikation und Transformation temporaler Anforderungen (Design of System Behaviour by Specification and Transformation of Temporal Requirements). *Proc. 18. GI–Jahrestagung II, Hamburg (R.Valk, ed.).* Informatik–Fachbericht IFB 188, Springer–Verlag, Berlin 1988, **449–463**.

[Rei85]   Reisig, W.: Petri Nets. Springer–Verlag, Berlin 1985.

[RiD82]   Richter, G., Durchholz, R.: IML–Inscribed Petri–Nets. *Proc. IFIP Work. Conf. on Comparative Review of Information Systems Design Methodologies (T.W.Olle, A.A.Verijn–Stuart, eds.).* North–Holland, Amsterdam 1982, **335–368**.

[Saa88]   Saake, G.: Spezifikation, Semantik und Überwachung von Objektlebensläufen in Datenbanken (Specification, Semantics and Monitoring of Object Life Cycles in Databases). *Doctoral Thesis.* Informatik–Skript Nr. 20, Techn. Universität Braunschweig 1988.

[Saa90]   Saake, G.: Descriptive Specification of Database Object Behaviour. Accepted for *Data & Knowledge Engineering*, Elsevier 1990.

[SaL89]   Saake, G., Lipeck, U.W.: Using Finite–Linear Temporal Logic for Specifying Database Dynamics. *Proc. CSL'88 2nd Workshop Computer Science Logic, Duisburg 1988 (E.Börger, H.Kleine Büning, M.M.Richter, eds.).* LNCS 385, Springer–Verlag 1989, **288-300**.

[Ser80]   Sernadas, A.: Temporal Aspects of Logical Procedure Definition. *Information Systems.* Vol. 5, 1980, **167–187**.

[SeS89]   Sernadas, C., Saake, G.: Formal Semantics of Object–Oriented Language Constructs for Conceptual Modeling. ISCORE Technical Report, INESC, Lisbon 1989, submitted for publication.

[SFNC84] Schiel, U., Furtado, A.L., Neuhold, E.J., Casanova, M.A.: Towards Multi-Level and Modular Conceptual Schema Specifications. *Information Systems.* Vol. 9, 1984, **43–57**.

[SFSE89]   Sernadas, A., Fiadeiro, J., Sernadas, C., Ehrich, H.-D.: The Basic Building Block of Information Systems. *Proc. Information System Concepts : An In-Depth Analysis (E.Falkenberg, P.Lindgreen, eds.).* North-Holland, Amsterdam 1989, **225-246**.

[SSE87]   Sernadas, A., Sernadas, C., Ehrich, H.-D.: Object-Oriented Specification of Databases : An Algebraic Approach. *Proc. 13th VLDB Int. Conf. on Very Large Databases. Brighton 1987 (P.M.Stocker, W.Kent, eds.).* Morgan Kaufmann Publ., Los Altos 1987, **107-116**.

[Wie90]   Wieringa, R.J.: Algebraic Foundations for Dynamic Conceptual Models. Ph.D. Thesis, Vrije Universiteit, Amsterdam 1990.

[WMW89] Wieringa, R.J., Meyer, J.-J., Weigand, H.: Specifying Dynamic and Deontic Integrity Constraints. *Data & Knowledge Engineering.* Vol. 4, No. 2, 1989, **157-190**.

[ZdM89]   Zdonik, S.B., Maier, D. (eds.): Readings in Object-Oriented Database Systems. Morgan Kaufmann Publ., San Mateo 1989.

Information Analysis: A step by step Clarification of Knowledge and Requirements

Helmut Thoma
CIBA-GEIGY Ltd.
Basle (Switzerland)

1. Abstract

Today, information analysis has two main topics: the construction of computer information systems and their integration. By integration, we mean not only sharing databases, but also other techniques for coupling applications. Using application architecture, we can design a rough picture of applications and databases for planning purposes for the whole or parts of an enterprise. We propose refinement of this basic picture for specific applications and databases producing process models and Entity-Relationship Models (ERM). Our ERM dialect can be transformed into relations into 4th Normal Form straightforward.

2. Introduction

Information analysis with its supporting methodologies is the most important task in modelling integrated computer information systems. Since the topic of integration has growing significance, chapter 3 stresses the aspects of integration when modelling information systems. The next four chapters deal with the process of modelling: collecting information requirements (chapter 4), identifying applications, databases, and the data exchange between applications (chapter 5), modelling the processes (chapter 6) and the data (chapter 7) of an application. In the final chapters, we discuss modelling approaches in general and some aspects of data management.

Throughout the paper, our framework is a schema (Figure 1), here called "Information Unit Pyramids". The left pyramid represents the functional aspects of information modelling, while the right one deals with data. The intersection shows a computer information system (or application) as an interaction of functions and data. From top to bottom the degree of detail, precision and formalization within each set increases.

In the following we focus on big companies with worldwide business. They employ EDP professionals in various organizational units and group companies. Organizational units and group companies normally take decisions on their own responsibility.

3. Objectives of Information Analysis

Information analysis can be defined, in the classical sense, as part of software engineering. Its primary objective is the construction of computer-supported information systems, also called applications. Through information analysis, the application designer clarifies and documents users' knowledge and requirements for data and the processes necessary to produce useful information. For several years, information analysis has relied on many different methods and tools for the design, development and implementation of computer information systems. Using information analysis processes through several development phases, an application model emerges which ideally can be used for generating the application. After a number of prototyping iterations, an application system should develop, integrating the knowledge and requirements of the users. In addition the results of the information analysis should be suitable as documentation for subsequent maintenance of the computer information system. The implemented system should be consistent with the application model.

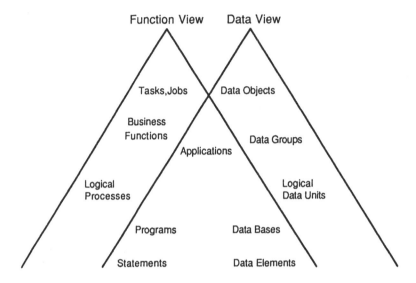

Fig. 1 Information Unit Pyramids

Comparing actual practice with the ideal often reveals discrepancies. These may be due to imperfect tools for the development process /Rieche et al 90/, for evolutionary prototyping, and for documentation compiled fragmentarily in individual phases of the development process. Other reasons are the frequent absence of a systematic approach to system development, short-term thinking on the part of those concerned with development projects, and misguided time and cost estimates. On the positive side, however, development methods are more or less well established.

A second objective of information analysis is improved application integration. In the past, information systems within the operational information environment have typically been isolated from each other. Today there are greater requirements for integrating isolated systems. Information resources have a more strategic impact in the marketplace and bring competitive advantages to the firm /McFarlan et al 83/. It is no secret that companies which are able to react flexibly to changes in the marketplace have more opportunities than those which don't.

"Just in time", as realized by implementation of concepts such as CIM /Scheer 87/ or CIB /Bullinger et al 87/ is an example of strategic objectives of this type. CIM (Computer Integrated Manufacturing) refers to the integration of economic functions such as production planning and control with technical functions such as Computer Aided Design and Computer Aided Manufacturing. Above CIM is a broader concept, CIB (Computer Integrated Business). From a technical and organization view, CIB includes all information and communication technologies in the enterprise.

Traditional application design and development, under which each application owns its data, is also not effective for decision support systems or ad hoc information retrieval. In order to realize the vision of "Executive Support Systems" or "Executive Information Systems", data and their definitions (semantics) must be available when and where they are needed.

With this in mind, a company should view individual "product pipelines" across the whole manufacturing line (value chain). However, today's information systems often focus on parts of the product-pipelines. Therefore it is essential that information permeate beyond the systems boundaries. Common access to shared databases would be the natural solution - if shared databases existed or were feasible.

At a minimum, data exchange should be supported between different information systems. Data exchange between applications at different sites or different companies, possibly with different hardware or software configurations, has an even greater importance. For this purpose there exist industry-wide standards or company-specific standards. On an industry basis, for example, international standards committees such as UN/EDIFACT have developed standards for Electronic Data Interchange. (For further information see /UN/EDIFACT 90/). Large enterprises, on the other hand, may reach internal consensus on application-specific standards for data exchange between sites or applications. However, the cost to the company of maintaining different data standards for (semantically) the same data is high. Manual consolidation is often necessary for this type of data exchange.

Application integration can therefore have two meanings. The first refers to the shared usage of the same physical data for different applications. The second refers to the computerized exchange of standardized data between different applications to assure unambiguous interpretation, requiring no manual intervention.

Before going into detail on that issue let us define some terms. By "Data of an Application" we mean that data is owned by that application. The term "Functional Dependency" is generalized: We say a database D is functionally dependent on a database S, if values in D are determined by values in S.

We distinguish four different degrees of application integration:

- Fully integrated: applications are fully integrated if they access the same physical data (shared database). Every application works at all times with the most current state of the database.

- Update-triggered: update to data in one application triggers the modification of correspondent data in a functionally-dependent database. Values in the functionally-dependent data base are not up to date (slight delay: The time for the proliferation of the modification).

- Update-triggered mailing: update to data in one application triggers a file transfer to a common mailbox. When necessary, the addressed application gets the data and updates its functionally-dependent database. The currency of this database depends on the last access to the mailbox.

- Time-triggered: at defined time intervals or at defined points of time updates to one database are copied to an equivalent functionally-dependent database. The currency of this database depends on the points of time of copying.

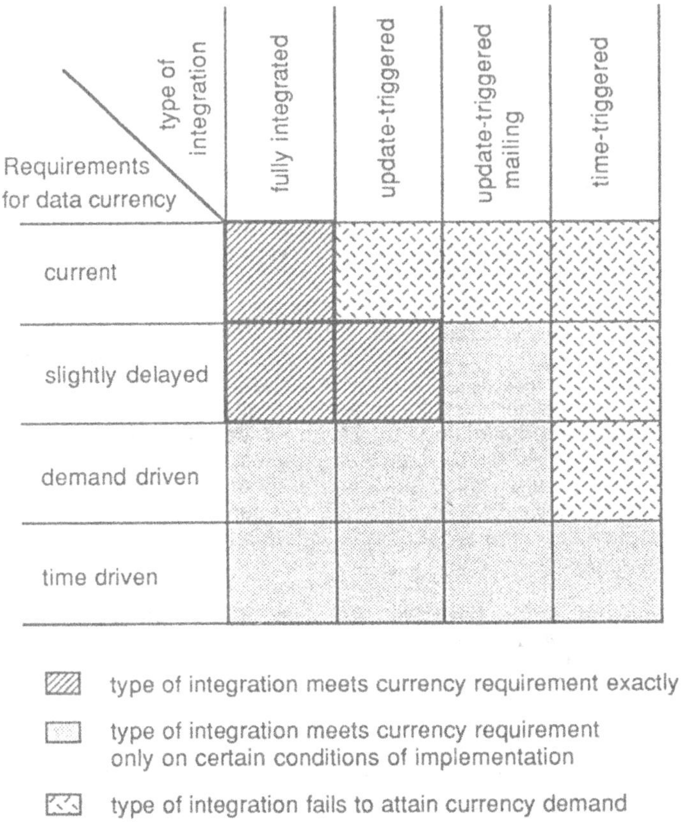

Requirements for data currency ╲ type of integration	fully integrated	update-triggered	update-triggered mailing	time-triggered
current				
slightly delayed				
demand driven				
time driven				

type of integration meets currency requirement exactly

type of integration meets currency requirement only on certain conditions of implementation

type of integration fails to attain currency demand

Fig.2 Requirements for data currency and related types of application integration

The last three degrees of integration operate with exchange of data in redundant databases. From now on, with "triggered integrated" we mean any of these three degrees.

The currency, or timeliness, of the output of an application is determind by the currency of its input database. The different degrees of application integration satisfy different data currency requirements.

A database that has to be current at any time must not be functionally dependent on a database of another application. Correct results are only obtained if the applications are fully integrated.

The currency of data of a functionally-dependent database is here called

- slightly delayed, if data are current with a few exceptions,

- demand driven, if the database can made current on demand (dynamically),

- time driven, if the database is made current at predefined points of time.

Figure 2 maps application integration to data currency. The figure shows that some degrees of integration

- meet some currency requirements exactly,

- meet other currency requirements under certain conditions like an appropriate flow of control of the applications or time stamped databases,

- fail to attain other currency demands.

Special attention must be given to data, triggered integrated with other data, that itself are triggered integrated with a third database etc. The compatibility of data currency requirements and the implementation of the total "triggered line" has to be considered.

So we conclude that information analysis must:

- Support conceptual application design,

- Support an approach for integrated applications to identify data integration requirements.

The following chapters present methods for systematic information analysis.

4. Collecting Information Requirements

This chapter presents an approach to identify the conceptual units (Tasks, Jobs, and Data Objects) from the two Information Unit Pyramids (Figure 1).

/De Antonellis et al 83/ present a three step approach to analyse the requirements for an application and to determine the scope of the "universe of discourse" through analysis of written texts. The products of this top-down process to develop an application down to the level of Logical Processes and Logical Data Units (Figure 1) are the "Data Glossary", "Operations Glossary" and "Events Glossary".

Their first step is to define the scope and requirements for the application, including documentation of current and planned functions in natural language and a list of organizational units involved in these functions. Then users are identified and listed who actually take part in some functions, who require information on the function or who can provide descriptions of how that function is performed within the organization. User interviews, questionnaires, business practices and procedures, and actual work observation are all viable requirement collection techniques. The Requirements Collection Form for each user of the function should contain a description of the function with detailed data about the creation, modification, verification and deletion of information objects.

Their second step is to filter and classify sentences. Filtering verifies that the texts are understandable and unambiguous and forces the use of language conventions. For example, filtering eliminates synonyms, repetitions and redundancy. Filtering requirements produces sentences referring to data, operations, events or constraints. It then extracts and collects them on specialized forms.

Their third step is to transcribe data, operations and events sentences into respective glossaries, according to their provenance. During the process and upon its completion, the designer must check that completeness and consistency rules are followed. The Data Glossary must contain properties for each data element. The Operations Glossary must reference all data for an operation. Each row of the Events Glossary must contain the data and operations belonging to an event.

Practical experiences with this approach in developing single applications are available in /Mayr et al 87/ and /Ortner et al 89/.

Selecting the right users to provide information requirements is extremely important. One must choose users who have a general view of the entire organization, with as little overlap as possible. This is especially important when the goal is to describe Tasks, Jobs and Data Objects with the approach outlined earlier. Users with too much knowledge or too much interest in details cloud the picture in early phases of integrated modelling.

In all cases one must define exactly the semantics of all information units.

This approach is nevertheless not adequate to satisfy our objective of designing more integrated applications. We can modify it by modelling according to the same method, but on a higher level of abstraction. For example, we can produce a first draft of Business Functions and Data Groups.

In order to link the functional view (Tasks and Jobs) with the data view (Data Objects), we take a different approach. Users tend to think in terms of organizational units and they know what information has to be exchanged between them. Therefore we construct a Directed Graph with organizational units as Vertices and the information which is exchanged between them as labelled Edges. The same facts can be presented in the form of the corresponding Adjacency Matrix with organizational units as rows and columns and pointers to flows of information as elements. Through discussions, interviews and evaluation of documents, we map activities of organizational units to Tasks and Jobs and information exchanged to Data Objects. The Tasks, Jobs and Data Objects feed into a first cut for Business Functions and Data Groups in the next step of creating an application architecture.

In performing this approach, we should question boundaries and activities of existing organizational units and think about new ones for future requirements. In any case we should center on duties and activities and not on organizational structures.

This approach is easier than the previous one because it provides a transparent integrated picture of data usage as a basis for discussion with the users.

5. Application Architecture

Developing an application architecture means identifying Business Functions and Data Groups by evaluating available information about Tasks, Jobs and Data Objects. It also requires clustering Business Functions and Data Groups according to their relations (refer to Figure 3). These clusters give an indication of the scope of applications and of global databases for shared use among applications.

5.1 Identify Business Functions and Data Groups

By combining related Tasks and Jobs from the requirements collection process, we develop a starting set of Business Functions. Combining related Data Objects forms the bases for Data Groups.

For each Business Function and Data Group, we need information about:

- Functionality and data of existing applications, whether they are developed inhouse or purchased;

- Functionality and data for planned systems;

- Strategic guidelines for future planning purposes.

The first step is to construct a table with Business Functions as rows and Data Groups as columns (see Figure 4). Through discussions with the users, the designer determines how each Business Function uses a Data Group, i. e. create, update or read. Create means that a Business Function can create or delete instances of at least part of a Data Group. Update means that a Business Function can create or change descriptive properties of a Data Group. Read means that a Business Function can read parts of a Data Group. Create implies update and read, while update includes read. For example, in Figure 4, Business Function B reads (R) data of Data Group 1, Business Function A creates (C) instances of Data Group 3 and Business Function C updates (U) properties of Data Group 1.

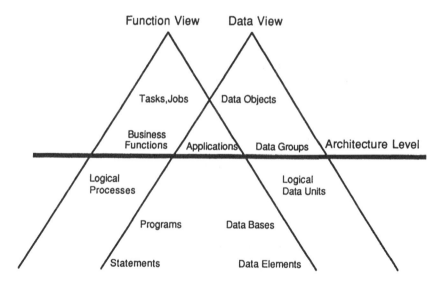

Fig. 3 Level of application architecture

Through the process of discussion and analysis, we reach with the users a deeper understanding of their data. It is especially revealing when several Business Functions appear to create or update the same Data Group. This may indicate the need to sub-divide a Data Group if differences exist in the data / data properties actually created / updated. The result may be to spread creates or updates over several columns.

A similar process can lead to refining Business Functions. A Business Function which is too broad will address too many Data Groups, and should be split into smaller units. The resulting Business Functions will each address fewer Data Groups.

Technical considerations such as an environment of multiple operating systems or database management systems can also suggest the need to split Business Functions or Data Groups. Organizational situations could also require the same type of splitting. An example would be a case where different organizational units perform the same Business Function.

It is essential to develop thorough documentation of all Business Functions and Data Groups as the analysis proceeds, to avoid future confusion or discrepancies. The information objects are not precisely defined, but users cannot provide both an overview and detailed information at the same time. Since we are emphasizing integration of information systems, we must concentrate on global information during the early development phases.

The process of clarification and refinement of Business Functions and Data Groups will be only successful by looking and discussing both in their interdependencies. Only in this way can the integration of data and applications be made in terms of use of data. A one-dimensional look at data or functions, for instance to get Business Functions or Data Groups at the identical level of detail, will fail in this context.

5.2 Identify Applications

We proceed to create a function and a data architecture to identify the applications needed for the business. The function architecture shows how Business Functions should be grouped to form an application. The data architecture combines Data Groups in pools according to similar rights of ownership.

The application definition process should answer the following questions:

- What data must be available in global databases for sharing by several applications (tightly coupled)?

- What data should be developed for which applications?

- What data should be stored redundantly?

- To what current data should each application have access?

- For what data is triggered integration of which applications sufficient?

- Are there existing applications which can handle any of the Business Functions, or are new applications necessary?

- How do we fit new applications into the existing environment, taking into account the data, functions and techniques we have available already?

First we identify existing and proposed applications and databases which we do not need to change, then mark the corresponding Business Functions and Data Groups. These marked items remain fixed throughout the subsequent process. However, if the boundaries of these applications and databases do not exactly match our Business Functions and Data Groups, some modification may be necessary (splitting).

Next we group Business Functions to applications. Organizational or technical constrains may affect the groupings. If this is not the case, grouping can be done by algorithm.

To use the algorithm approach, we calculate affinity coefficients between pairs of Business Functions by considering the number of jointly used Data Groups. Then we join those Business Functions with affinity coefficients exceeding a certain threshold. This iterative process, here called "joining by superposition", continues until all affinity coefficients fall below the threshold. The affinity coefficients calculation should weight the various Data Group usage categories differently for create, read, update. The clustering results should be reviewed carefully for business plausibility.

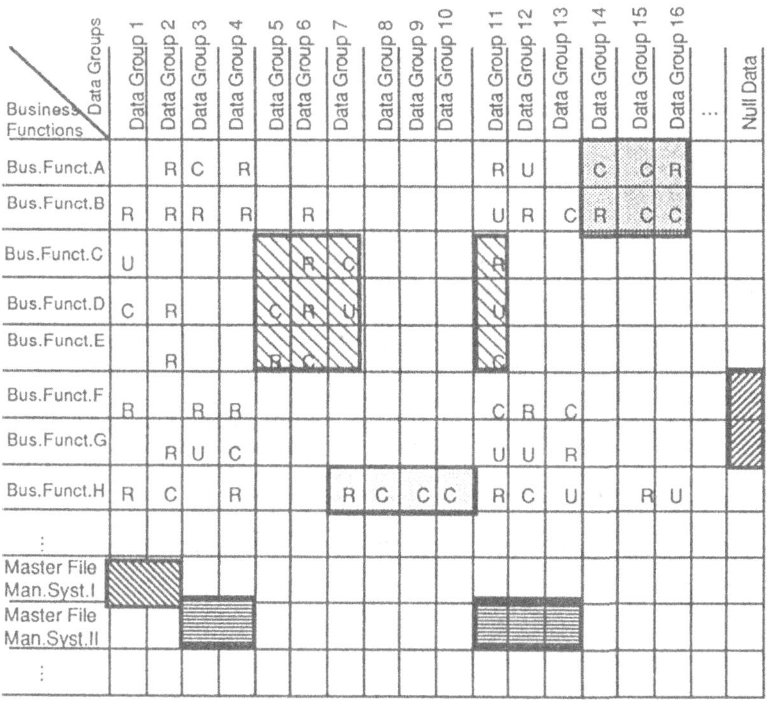

Business Functions \ Data Groups	Data Group 1	Data Group 2	Data Group 3	Data Group 4	Data Group 5	Data Group 6	Data Group 7	Data Group 8	Data Group 9	Data Group 10	Data Group 11	Data Group 12	Data Group 13	Data Group 14	Data Group 15	Data Group 16	...	Null Data
Bus.Funct.A		R	C	R							R	U		C	C	R		
Bus.Funct.B	R	R	R	R		R					U	R	C	R	C	C		
Bus.Funct.C	U				R	C					R							
Bus.Funct.D	C	R			C	R	U				U							
Bus.Funct.E		R			R	C					C							
Bus.Funct.F	R		R	R							C	R	C					
Bus.Funct.G		R	U	C							U	U	R					
Bus.Funct.H	R	C		R			R	C	C	C	R	C	U	R	U			
Master File Man.Syst.I																		
Master File Man.Syst.II																		

Fig. 4 Application architecture chart

The clustering of Data Groups into pools and their assignment to applications results from the local importance of every Data Group for one or a number of applications and from further important additional conditions such as implementation technology. Different implementation techniques can lead to assignment of a Data Group to more than one pool. This leads to data redundancy and requires further measures to ensure consistency.

Data Groups with predominant local importance for any one application are clustered in one application-specific pool with application-specific control, i.e. only one application creates and updates the data.

Data Groups important for more than one application are combined in global databases with their own master file management systems.These data are available to all applications. If one application uses data owned by another application or by a master file management system, this implies integrated, shared usage of data. Applications without their own data arise primarily through the creation of global database applications.

In Figure 4, three applications with application-specific pools of data are shown:

- An application for Business Functions A and B with Data Groups 14, 15 and 16,

- An application for Business Functions C, D and E with Data Groups 5, 6, 7 and 11,

- An application for Business Function H with Data Groups 7, 8, 9 and 10.

Further, we see two pools of global data, administered by master file management systems. The data pools belong to no specific Business Functions because they will be of general use. One pool contains Data Groups 1 and 2, the other pool contains Data Groups 3, 4, 11, 12 and 13.

Also we have an application which owns no data and accesses only data owned by other applications or master file management systems. This application includes Business Functions F and G and is marked in the column "Null Data".

We see that data from Data Group 7 are redundant (a copy of Data Group 7 is maintained by the application for Business Function H). All applications access data of master file management system I (Data Groups 1 and 2). The application for Business Functions C, D and E does not directly access the data of master file management system II (Data Groups 3, 4, 11, 12 and 13), but the other applications do.

From an application architecture chart we can derive the following:

- An integrated data model - with less detail than an entity-relationship model,

- Applications,

- Shared and application-specific databases,

- Redundant data and replicates for distributed data,

- Data transfer requirements and interfaces for direct access.

The architecture is useful for studying the consequence of different technical environments for implementing and distributing the applications and databases. Standard software can also be integrated into the application architecture. In addition to modelling the past and present, the application architecture can function as a planning tool by incorporating future requirements.

Several facts are still to be documented outside of the application architecture chart. Thus, the chart shows only if the type of integration is fully integrated or only triggered integrated, but not the exact degree of triggered integration. And the requirements for data currency must also be documented elsewhere.

For which sectors of an enterprise does an application architecture make sense?

In big companies, we might focus on a large organizational unit, such as a division. From the point of view of the entire company, this is only partial integration. An alternative is to concentrate on the most important data from a total company perspective. This approach requires bringing together representatives from each organizational unit, and may be impractical simply because of logistics.

When modelling data across organizational boundaries, for example among a group of relatively independent group companies, the best solution is to model and standardize on a small set of data important to all of the business units. It is useful to consider existing technology and definitions such as the UN/EDIFACT standards for electronic data interchange.

Existing published methods and tools for application architecture design are not quite satisfactory. Using the "Information System Study" (ISS), IBM offers a systematic approach to form applications as clusters of functions and data so that data exchange between applications becomes minimal. A tool called ISMOD is useful for this approach. (/Hein 85/ describes ISS and ISMOD in detail.) In our opinion, the results of this method are not satisfactory because it emphasizes isolated applications and the transfer of data between them.

/Martin 83/ describes another method which leads to the formation of optimal data units by using an affinity matrix for data aggregation. In this approach, the frequency of commonly used data is very important. Martin concentrates on consolidating commonly used data into global databases, or "subject databases". He does not deal with forming applications at all. In three projects we used the Martin clustering algorithm, implemented in IEW's planning workbench, but we created initial proposals for applications, not for databases as described by Martin.

Both the ISS and Martin approaches do not deal with the question of incorporating standard software.

An approach to planning applications with similar objectives as our method of application architecture is presented in /Brancheau et al 89/. It is also similar in its perception of functional and data aspects. Their evaluation, however, considers mainly business-oriented information requirements. It does not group Business Functions and Data Groups according to their mutual interdependence and it does not consider technical aspects.

6. Process models

Process models should support a refined description of Business Functions. Their purpose is to clarify the computer-aided parts of the function view of applications, i. e. the set of Business Functions belonging to one application.

Graphical process models - often used in Software Engineering Tools - are based on hierarchical structures of processes, subprocesses etc. or on the flow of control between processes and between subprocesses. Process models are frequently produced by stepwise refinement of process units ("Logical Processes") and are influenced by Petri-Nets.

A process model can be elaborated to one graph per process with operations as vertices. Figure 5 is an example of a model for one process. Event A and Event B must exist as input conditions for Operation 1. Under Condition 1-C1, Operation 1 produces output D, and under Condition 1-C2, Operation 1 produces output E. Output D or output C make Operation 2 work. Outputs H and I are results of this small process model and may be inputs to another process model.

An event is an external input such as a time pulse or the output of another process. A result is an external output of an operation and may be an input to another process.

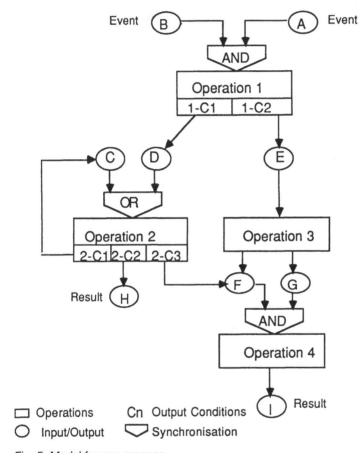

Fig. 5 Model for one process

7. Data models

Data modelling techniques refine the structure and description of Data Groups. Normally the boundaries of one data model are roughly determined by an application or by a master file management system. If we model without a CASE tool, our notation is a dialect of the Entity-Relationship Model (ERM), discussed in Chapter 7.1. The ERM was introduced by /Chen 76/, many derivatives have since appeared. A new textbook for students is /Elmasri et al 89/, providing an important conceptual understanding of data.

The characteristics of our ERM notation are

- Utility for communication between analysts and users,

- Balance between semantical completness and syntactical simlicity,

- Ease of transformation into relations in 4th Normal Form.

We consider data modelling to be more important than process modelling, because only data-oriented design leads to information systems that can be integrated. Furthermore, data modelling is more useful than process modelling for integrating knowledge-based systems with conventional systems.

An ERM is easy to understand for users when analyzing Data Groups, since it concentrates on the semantics, or descriptions, of data.

When modelling Data Groups, we look at the needs of all Business Functions involved. The data models must reference all known usages for the data so that we can prepare for integrated processing. We do not analyze access paths, however, until physical database design, when performance becomes an issue. Here we deal only with modelling the logical dependencies of the data elements.

On the Information Unit Pyramids (Figure 1), we will now focus on Logical Data Units and Data Bases. For relational database management systems, we include databases in the form of base tables.

7.1 Entity-Relationship Model

As we identify units of Data Groups, we refine them with the Entity-Relationship Model. This is a collaborative effort with users. In critical phases, a neutral expert should facilitate the modelling process.

Objects, individuals, events or concepts can be entities and are condensed into entity types, modelling corresponding objects, individuals etc. as a set of entities in the sense of Cantor's set theory. In Figure 6, the entity types are "Department" and "Employee".

One entity contains at least one property, namely the one which unambiguously identifies that entity. The appropriate property type is documented as an identifier. In Figure 6, "Dept.#" is the identifier for entity type "Department".

There can be relationship types between entity types which may be simple or complex, depending on how many entities of one type may be related to a given number of entities of the other type. We distinguish 1:1 (or one-to-one), 1:n (or one-to-many), and n:m (or many-to many). The first one is rarely used.

In Figure 6, one department, identified by "Dept.#", can have more than one employee. One employee, identified by "Empl.#", may belong to only one department.

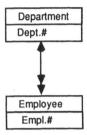

Example of an (1:n) relationship type between the entity types "Department" and "Employee"

Fig. 6

A double-sided complex relationship type between entitiy types requires special modelling. This type of relationship leads to the formation of an (n:m) relationship type as shown in Figures 7 and 8. Several entity types, (n:m) relationship types, or entity types and (n:m) relationship types can be related by one (n:m) relationship type.

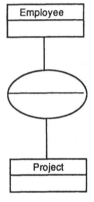

Example of a (n:m) relationship type between the two entity types "Employee" and "Project".

Fig. 7

Figure 7 shows that one employee can work on several projects, and one project can involve several employees.

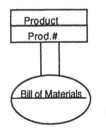

Example of a recursive (n:m) relationship type within the entity type "Product", which is labelled with "Bill of Materials" .

Fig. 8

Figure 8 shows the "Bill of Materials", a recursive (n:m) relationship type, where one product can consist of several components, or intermediate products, and one component may be required to produce several products. (A component also is a product.)

Like entity types, (n:m) relationship types must have a name and an identifier for each relationship which is a combination of the identifiers of the entity types or (n:m) relationship types involved in the relationship (see Figures 10 and 11).

There exist also recursive (1:n) relationship types. For example, modelling the hierarchical structure in an organization, one employee has only one immediate manager, who himself is an employee, whereas one manager could have more than one subordinate.

As earlier discussed, entities within entity types and relationships within (n:m) relationship types both possess at least one property - a unique identifier. Entities and (n:m) relationships may have other properties. If they do, properties, describing corresponding facts, are condensed into one property type. Property types can exist only in connection with at least one entity type or (n:m) relationship type. This dependence, which may be simple or complex, is shown as arrows beween property types and entity types, (n:m) relationship types or other property types.

Figure 9 shows that an employee has only one name and one primary residence. Several employees, however, may have the same name or the same main residence. The employee's residence determines the country of tax obligation.

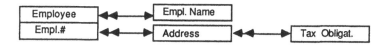

Fig. 9 Property types "Empl. Name", "Address", and "Tax Obligat." and
their dependencies

Property types are inferred from their corresponding entity types or (n:m) relationship types. There is no need to label a dependency arrow if the meaning of the property type is clear from the context and names. A label (role, characteristic) is necessary when different dependencies exist for the same property type. The equivalent remarks are valid also for (1:n) relationship arrows and for the edges of (n:m) relationship types, as shown in the Bill of Materials example in Figure 10.

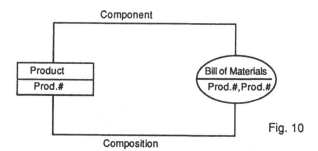

Fig. 10

For readability and understanding of our model, we should put as much information as necessary and as little as possible into the graph. Since there in practice may be easily up to 200 property types for one entity type or (n:m) relationship type, we should combine property types having the same kind of dependency into one form, then reference this form in the Entity-Relationship Model.

We develop our ERM in several phases and iterations. First we should focus on the entity types and relationship types to form a base version of the ERM. If necessary, the neutral moderator may verify and amend the ERM with users and system developers. Identifiers are important wherever they are needed for clarity and understanding. In the second phase, we determine property types, their dependencies and, where necessary, labels.

In practice, we found that the upper limits on instances in a relationship between entities or in a dependence, as denoted by arrows, are not sufficient. We often need a way to specify lower limits, and can use the notation in Figure 11. For example, this figure shows that a project need not have a name, but must have one (and only one) indication of manpower. Later, in the relational data model, these lower limits are very important to identify null values and the constraints of referential integrity.

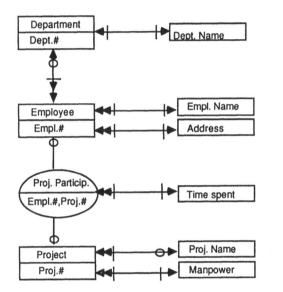

Fig. 11

The ERM in Figure 11 summarizes some constructs discussed above. It describes the set of employees within the entity type "Employee", its identifier is "Empl.#". An employee has one and only one name, "Empl. Name" and one and only one address, "Address" (descriptive properties). The name and address belong at least to one employee and may belong to more than one employee.

The entity types "Department" and "Project" model departments and projects. A department has the identifier "Dept.#" and a project has the identifier "Proj.#". A department has one and only one name ("Dept. Name"), which applies to one and only one department. In this case "Dept. Name" is an alternate identifier of "Department". A project does not have to have a name ("Proj. Name"), but if it does, the name can be valid for more than one project. Each project has one and only one manpower cost associated with it, but several projects may have the same cost.

Now we will discuss the relationships between the entities. An employee can belong to only one department, but does not have to belong to a department. One department can have many employees (a (1:n) relationship type). One employee may work on several projects (between 0 and n), while one project may have several employees working on it (an (m:n) relationship type). This relationship type, called "Proj. Particip." has an identifier for determining the instances. In our example, this is the combination of "Empl.#" and "Proj.#", hence "Empl.#, Proj.#". For each project and each employee, the manpower used is called "Time spent", which may be the same for more than one project and one employee.

Many CASE tools do not allow for modelling of (n:m) relationships (Figure 12). In this case, the construct must be replaced by a new entity type and two (1:n) relationship types, as shown in Figure 13. The two constructs, however, are not equivalent. In Figure 12, the identifier of sales unit is built by combining identifiers of the related entity types. This guarantees that a single combination of "Prod.#" and "Pack.#" can exist only once. Figure 13 shows a sales unit identifier, "SU#", with its own existence. Here, "Prod.#" and "Pack.#" are foreign keys, to use relational terms. One combination of these two can exist many times.

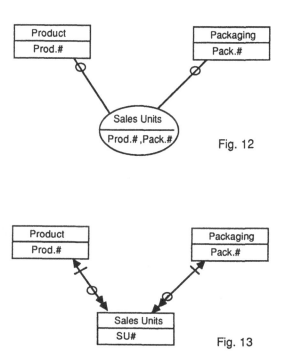

Fig. 12

Fig. 13

Figure 14 shows our construct for generalization and specialization. In our example, a business partner may be a customer or a supplier or both. For this reality two "is-a"-(1:1) relationship types are created, joined with a logical "OR"-operator. For customers and suppliers, the identifiers are marked with an asterix (*), if we use for specialized entity types identifiers from the same domain as for the generalized entity type. In the other case we use separate identifiers for specialized entity types (Figure 15). Corresponding to "OR" we may also use "XOR" (exclusive or) where appropriate.

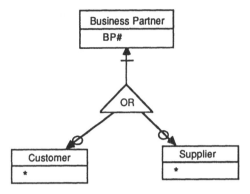

Fig. 14   Construct for generalization: Identifiers for the generalized and
and the specialized entity types are the same

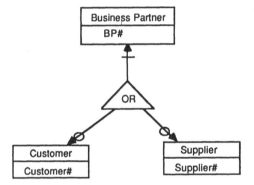

Fig. 15   Construct for generalization: Identifiers of the generalized and the
specialized entity types differ

7.2 Transformation into a Relational Data Model

The Entity-Relationship Model is the basis for the relational data model. Relations and attributes are defined from the entity types, relationship types, and property types. The following rules determine transformation of the ERM to relations in 4th Normal Form. (Note: in this rules, property type also means identifier).

The first two transformation rules are usually adequate. The remaining rules are used only when entity types or relationship types are not recognized as such.

1.   Each entity type and each (n:m) relationship type defines a relation with the identifier as primary key.

2.   The names of the property types are added to the relations as attribute names, if the property types can directly be reached by a simple relationship or depen-

dency arrow from the particular entity type or (n:m) relationship type (the identifier of a neighbouring entity type or (m:n) relationship type will be a foreign key).

Should this dependency be (1:1), then this attribute is an alternate key. Should the relationship be (1:1), then the name of an identifier must be added in only one direction, to avoid redundancy. If the dependency or the relationship arrows are labeled, the name of the attribute will be the concatenation of the label of the arrow with the name of the property type.

3. When property types contain additional property types in a simple dependency (transitive dependency), we define an additional relation. The name of the property type from which the simple dependency starts becomes the primary key of the new relation.

4. When an (n:m) dependency exists between an entity type or (n:m) relationship type and a property type or between property types, we define additional relations. Each property type involved becomes a relation, with primary key equal to the name of the property type. Each (n:m) dependency also becomes a relation whose primary key is a combination of the names of both related property types.

5. A property type becomes a new relation if it is the source of a simple dependency on an entity type or a (n:m) relationship type, upon which it complex depends. The name of the property type becomes the primary key. The equivalent trans-formation must be done if the property type is the source of a simple dependency on another property type.

6. For relations created according to rules 3, 4 and 5, the corresponding property types must be added as attributes.

Examples of transforming Entity-Relationship Models from Chapter 7.1 into relations in 4th Normal Form follow:

Transformation of Figure 9:

Employee (Empl.#, Empl. Name, Address)
Tax (Address, Tax Obligat.)

Transformation of Figure 10:

Product (Prod.#)
Bill of Materials (Prod.#. Component, Prod.#. Composition)

Transformation of Figure 11:

Department (Dept.#, Dept. Name)
Employee (Empl.#, Dept.#, Empl. Name, Address)
Project (Proj.#, Proj. Name, Manpower)
Proj. Particip. (Empl.#, Proj.#, Time spent)

<u>Transformation of Figure 12:</u>

Product (<u>Prod.#</u>)
Packaging (<u>Pack.#</u>)
Sales Units (<u>Prod.#</u>, <u>Pack.#</u>)

<u>Transformation of Figure 13:</u>

Product (<u>Prod.#</u>)
Packaging (<u>Pack.#</u>)
Sales Units (<u>SU#</u>, Prod.#, Pack.#)

<u>Transformation of Figure 14:</u>

Business Partner (<u>BP#</u>, < general attributes for business partners>)
Customer (<u>BP#</u>, < specific attributes for customers>)
Supplier (<u>BP#</u>, < specific attributes for suppliers>)

<u>Transformation of Figure 15:</u>

Business Partner (<u>BP#</u>, < general attributes for business partners>)
Customer (<u>Customer#</u>, BP#, < specific attributes for customers>)
Supplier (<u>Supplier#</u>, BP#, < specific attributes for suppliers>)

For the implementation of the relations, it is necessary to transform the relation names and the attribute names into specific internal database names.

8. Discussion of design approaches

Our approach emphasizes data orientation in context to the functional usage of data and top down integration in planning an application architecture. We use only slightly detailed information units. In early phases of design, however, the user does not yet has detailed knowledge. Analysis in depth makes sense when we are developing a specific application or a specific global master file and have to write down detailed models of Logical Processes and Logical Data Units.

Other methods are also in practical use. Historically, most developers used function-oriented methods. In these methods, one first defines the processes of the particular application, then identifies and structures data for optimal access (practical example: about 1300 IMS databases). Database management systems were often looked at as reliable (recovery) and nevertheless fast file systems.This methodology results in "data chaos": excessive data redundancy and applications which remain isolated from each other /Boulanger et al 89/.

By shifting our emphasis from functions to data, for example with Entity-Relationship Modelling, we produce stable data models, but still face the application isolation problem.

In our opinion bottom-up integration by itself cannot solve these problems either. By bottom-up integration we mean the construction of an overall data model by stepwise integration of application specific data models. As a base we could use existing components of models of similar parts of the real world. Recognition of the meaning of components of existing models and especially to perceive the meaning (semantics) of data elements in historically grown data (practical example: more than 50000 data elements) is only possible with exorbitant effort: in the past, data definition has seldom been done with tools which allow subsequent semantical analysis of the definitions (see also /Brenner 88/). Thus it is nearly impossible to locate redundancy of data.

Even if it were possible to integrate existing data models using common data elements, this way to fully integrated applications is very expensive and tedious. In a practical case the estimated man power for the migration of one application-specific IMS database to a global master file was 200 man months. The only way to migrate data of an old architecture into new data of a new architecture will be a stepwise procedure in several phases over years /Data Architecture 89/. A helpful approach to integrated usage of existing data could be the implementation of a global database, time triggered integrated with databases of operative online applications. This temporary solution has been realized by IBM in Mainz under the name "Data Warehouse" /Brendel 90/.

9. Data Management

To benefit from the results of information analysis during the whole life cycle of applications and to document standards for shared and triggered data, data management functions must be organized /Holloway 88/.

The most important data management tool is a data dictionary system (repository etc.). This should document meta data, the formalized and formatted descriptions of all parts of the Information Unit Pyramids. There is rarely a controversy over the documentation of physical units used for schema generation - the dispute usually centers around documentation of logical units. Without the computer-assisted documentation of logical units, especially at the basic application architecture level, there is little opportunity for application integration /Thoma 88/.

The main use of data dictionaries is in the system development and maintenance environment. To assure commonality of thought and procedures, standardizing the most important data is crucial. Beyond this, we propose standardizing on all data used in more than one application or with global scope. The simplest documentation format for standards are attributes and relations in 4th Normal Form.

10. Acknowledgement

I would like to thank Paula Schutte and Hans Peter Uebersax for reviewing the paper and contributing constructive criticism. I also wish to acknowledge Pamela Vogelbacher, Nicholas Palmer and Tom Treuer for their help in the editorial work and my managers Günther Burris and Walter Schelldorfer for their positive attitude.

11. References

/Boulanger et al 89/
>   Boulanger, D.; March, S. T.: An approach to analyzing the information content of existing Databases. Data Base 20 (2), 1-8 (1989)

/Brancheau et al 89/
        Brancheau, J. C.; Schuster, L.; March, S. T.: Building and Implementing an
        Information Architecture. Data Base 20(2), 9-17 (1989)

/Brendel 90/
        Brendel, M.: CIM im IBM-Werk Mainz. Presentation, 1990.

/Brenner 88/
        Brenner, W.: Entwurf betrieblicher Datenelemente. Reihe "Betriebs- und
        Wirtschaftsinformatik", Band 28. Berlin, Heidelberg: Springer. 1988.

/Bullinger et al 87/
        Bullinger, H.-J.; Fähnrich K.-P.; Thines M.: Einbettung von
        CIM- Konzepten in ein unternehmensweites Informations-
        management. In: Bullinger, H.-J. (Hrsg.): Tagungsband
        kommtech 87, Kongress III, Computer Integrated Manufacturing
        und Unternehmenslogistik. Velbert: ONLINE GmbH. 1987

/Chen 76/
        Chen, P. P.-S.: The Entity-Relationship Model - Toward a unified View of
        Data. ACM Transactions on Database Systems 1 (1), 9-36 (1976)

/Data Architecture 89/
        Give your organization new information capabilities. Data Architecture,
        Vol. 1, Nr. 3. Atlanta: Data Modelling Group. 1989

/De Antonellis et al 83/
        De Antonellis, V.; Demo, B.: Requirements Collection and Analysis. In:
        Ceri, S. (Ed.): Methodology and Tools for Data Base Design. Amsterdam:
        North-Holland. 1983.

/Elmasri et al 89/
        Elmasri, R.; Navathe, S. B.: Fundamentals of database systems. Redwood
        City: Benjamin / Cummings. 1989

/Hein 85/
        Hein, K. P.: Information System Model and Architecture Generator. IBM
        Systems Journal 24 (3/4), 213-235 (1985)

/Holloway 88/
        Holloway, S,: Data Administration. Hants: Gower Technical Press. 1988.

/Martin 83/
        Martin, J.: Managing the Data-Base Environment. London: Prentice-Hall.
        1983

/Mayr et al 87/
        Mayr, H. C.; Dittrich, K. R.; Lockemann, P. C.: Datenbankentwurf. In:
        Lockemann, P. C.; Schmidt, J. W. (Hrsg.): Datenbank-Handbuch. Berlin,
        Heidelberg: Springer. 1987.

/McFarlan et al 83/
        McFarlan, F. W.; McKenney, J. L.: Corporate Information
        Systems Management. Homewood: Richard D. Irwin. 1983.

/Ortner et al 89/
    Ortner, E; Söllner, B.: Semantische Datenmodellierung nach der
    Objekttypenmethode. Informatik-Spektrum 12 (1), 31-42 (1989)

/Rieche et al 90/
    Rieche, H. J.; Thoma, H.: Hilfsmittel für die Systementwicklung: Erwartung
    und Realität - ein Erfahrungsbericht. Computer Magazin 19 (3/4), 49-54
    (1990)

/Scheer 87/
    Scheer, A.-W.: Computer Integrated Manufacturing: CIM =
    Der computergesteuerte Industriebetrieb.Berlin, Heidelberg:
    Springer. 1987.

/Thoma 88/
    Thoma, H.: Die Rolle des Data Dictionary bei der Systemintegration. In:
    Oertly, F. (Hrsg.): Data Dictionaries und Entwicklungswerkzeuge für
    Datenbank-Anwendungen. Tagungsband DBTA/SI. Zürich: Verlag der
    Fachvereine. 1988.

/UN/EDIFACT 90/
    UN/EDIFACT Rapporteurs´ Teams: Introduction to UN/EDIFACT with
    latest News and Events. United Nations / ECE. April 1990

Model-Based Knowledge Acquisition

Angi Voß
Expert Systems Research Group
GMD
Postfach 1240
D-5205 Sankt Augustin

e-mail: avoss@gmdzi.uucp

Abstract

There are two knowledge engineering traditions, rapid prototyping and conceptual modelling which can further be distinguished into universal and shell-based approaches. Underlying the two traditions are two different hypotheses explaining the knowledge acquisition bottleneck. Model-based approaches try to tackle it by searching for a set of basic, generic problem solving methods that can be combined and instantiated for various applications. KADS is introduced as a universal conceptual modelling approach and it is compared with three prominent shell-based ones. Summarizing, knowledge engineering will be compared to software engineering which will suggest to view knowledge acquisition as a workbench of methods and techniques that are special to the former.

The paper will close with the insights I gained on the workshop about the relevance of knowledge engineering for data base engineering, and vice versa.

1   Two traditions in knowledge engineering

Knowledge Engineering denotes the discipline and the process of operationalising expertise on the computer. Having emerged from Artificial Intelligence, *rapid prototyping* or evolutionary programming approaches were initially prevailing. As soon as possible one started to build a small system which was successively refined until a satisfactory version had been obtained. If the prototype turned out to be inadequate it was discarded and a completely new system was built. At first, knowledge-based systems were directly programmed in Lisp, Prolog, in a rule-based formalism or an object-oriented programming language [Hayes-Roth, Waterman, Lenat 83]. Since these languages are universal, I will call such approaches *universal prototyping*.

With increasing experience, one began to extract the problem solving procedures from specific systems, generalized them and wrapped them into shells [Puppe 86], [Cunis, Guenter, Syska 87]. This was a significant improvement. If you were lucky enough, there was a suitable shell for your problem, so that you just had to fill in the application-specific knowledge. But time and again, the luck is being forced and shells are misused. I will use the term *shell-based prototyping* for approaches using such special purpose shells.

Applications growing more complex, the level of universal languages was felt to be inadequate. The same was true for shells, insofar as they could be understood only in terms of their implementation languages. Higher level, problem specific, conceptual descriptions were wanted. Naturally, these were first obtained by abstracting the procedures employed in the shells to *problem solving methods* or *generic conceptual models* [Clancey 85], [Chandrasekaran 86], [McDermott 88], [Musen 89].

However, due to their different provenience, they differed in their vocabulary and were not adaptable or combinable.

The *KADS* group then proposed a scheme and a vocabulary for describing arbitrary models, i.e. problem solving methods and their applications to a particular domain, on the conceptual level [Wielinga, Breuker 86], [Breuker, Wielinga 89], [Schreiber et al. 87]. The vocabulary being semiformal with an intuitively given semantics, these models were not directly executable. They were intended as specifications which are to be followed by a so-called *design-model* and an implementation. Due to their semi-formal nature, KADS models could be used to describe problem solving behaviour as encountered in the world, which even might not be operationalizable entirely.

By now, model-based knowledge engineering involves developing or finding a conceptual model. If the model is the abstraction of a shell, one has to fill in the application-specific knowledge as in shell-based prototyping, otherwise, the model can only be used as a specification. In analogy to the rapid prototyping approaches, I will use the terms *shell-based* and *universal modelling approaches* depending on whether the models are shell abstractions or can be freely defined. Figure 1 illustrates the descriptions obtained by the different approaches and figure 2 summarizes their properties.

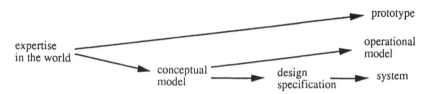

Figure 1: Descriptions produced by rapid prototyping and modelling approaches

approach	operational	universal	conceptual level
universal rapid prototyping	X	X	
shell-based rapid prototyping	X		
shell-based modelling	X		X
universal modelling		X	X
model-based prototyping	X	X	X

Figure 2: Advantages and disadvantages of different types of approaches

As its major advantage, the KADS approach allows to invent new models, or to adapt and combine existing ones. This is necessary since we are far away from a complete library of problem solving methods, and since every real-life application will usually combine several of them. However, KADS models cannot be tested, which seems to be a sine qua non for complex applications. You just cannot conceive of all possible interactions between all components of your model. On the other hand, early testing was the advantage of rapid prototyping.

Thus, I personnally favour an approach that is referred to as *model-based prototyping* in the last line in figure 2. It aims at operationalizing conceptual models as they grow, so that they can be experimented with as soon as possible [Karbach, Tong, Voss 88], [Voss et al. 90]. The resulting complete operational model may be the endproduct or serve as a specification. A re-implementation may be necessary for reasons of efficiency, but it should preserve the structure of the model as a conceptually adequate description. And conceptually adequate systems are easier to understand, explain, debug and maintain.

In the next section I will describe the problem referred to as the knowledge acquisition bottleneck and explain how the two knowledge engieering traditions explain it. How model-based approaches try to cope with it will be the subject of the following sections. By then, the reader should be in a better position to judge my comparison between knowledge engineering and conventional software engineering, and the resulting definition of knowledge acquisition in section 6.

2   The knowledge acquisition bottleneck

Various decompositions of knowledge engineering into phases and subtasks have been proposed, however, a unique view has not yet emerged [Karbach, Linster 90]. The term *knowledge acquisition* has been used synonymously to knowledge engineering, but also for one of its phases or subtasks. Similarly, the term *knowledge elicitation* is sometimes used identically with knowledge acquisition, sometimes for one of its phases or subtasks. As I will explain in section 6, I prefer to view knowledge acquisition as a collection of methodologies, methods, techniques and tools, while I consider knowledge elicitation as a subset thereof.

Together with [Ueberreiter 90], I (currently) prefer to view knowledge acquisition as decomposed into three tasks: eliciation, structuring and operationalization. *Elicitation* deals with extracting knowledge about problem solving from persons or documents. The result may be various, often informal documents. *Structuring* means to interpret the elicited knowledge and to develop a problem solving method that can eventually be realized on a machine. Elicitation must precede structuring, but it is also necessary to fill in the contents of the structure, yielding a complete model. *Operationalization* means to realize the model on the machine, which can be done directly or via a new design. Operationalization thus subsumes knowledge representation or shell selection.

Knowledge engineers soon recognized that eliciting knowledge and transfering it into a program is not easy at all. This phenomenon was called the *knowledge acquisition bottleneck*. In the rapid prototyping tradition, it was explained by what I will call the *mapping hypothesis*: Knowledge elicitation techniques were believed to be insufficient. Better techniques would allow to dig deeper so as to produce enough knowledge to be transferred into the prototype.

By now, this hypothesis has been falsified. Cognitive scientists distinguish three major categories of knowledge,which are supposed to be traversed sequentially in the process of becoming an expert. *Explicit knowledge* is told explicitly and is accessed consciously. With growing experience, it turns into *associative knowledge* and finally is compiled into *tacit knowledge* which is no longer accessible by will [Musen 89]. Figure 3 shows that most elicitation techniques that can be practicably applied in knowledge engineering cover only conscious knowledge and partially associative knowledge, while tacit knowledge is practically inaccessible, at least by techniques that can be used during knowledge engineering.

Consequently the expert's problem solving behaviour cannot be mapped or simply transferred, but a problem solving method must be (re-) constructed. And even if the experts happen to know how

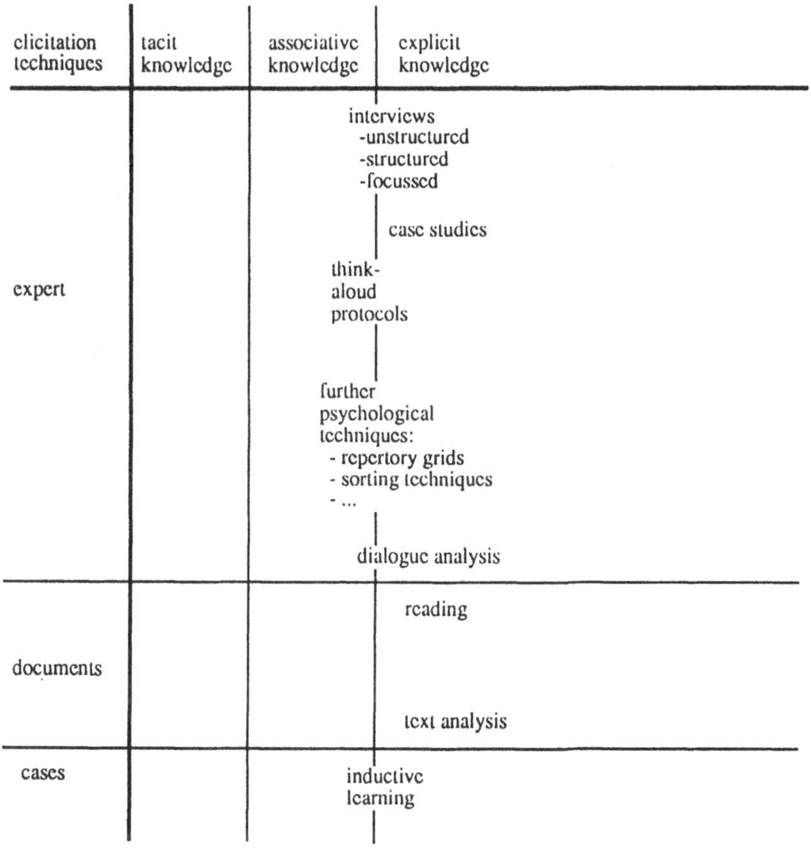

elicitation techniques	tacit knowledge	associative knowledge	explicit knowledge
expert		interviews -unstructured -structured -focussed case studies think- aloud protocols further psychological techniques: - repertory grids - sorting techniques - ... dialogue analysis	
documents			reading text analysis
cases		inductive learning	

Figure 3: Elicitation techniques that qualify for knowledge engineering and the categories of knowledge they may elicit

they do it and can verbalize it, their method may not be transferrable to the machine, or it might be too inefficient.

The problem solving method to be (re-)constructed should be documented somehow. In rapid prototyping approaches, the first document is the prototype itself. This is rather unsatisfactory, since it is on too low a level involving a lot of implementation details. A higher level description is wanted, in particular, since the solution is to be elaborated and agreed upon by experts, users, knowledge engineers and programmers together. That means, knowledge engineering has been recognized as a social process, and conceptual models were introduced as a high level, conceptual specification that can be discussed by all persons involved or affected. That is why [Johnson 88] calls it a *mediating representation*.

Thus, the problem of digging out all the expertise has been shifted to the problem of (re)inventing it. To tackle this one, the model-based tradition relies on another hypothesis, which I will call the *assembly hypothesis*: It is hoped that there is a finite and small set of generic problem solving methods, such that every expert problem solving behaviour can be realized by a composition thereof. In order to evaluate this hypothesis, classifications of problems and problem solving methods are being constructed and criteria are being developed for finding the right method combination for a given problem.

3 Classification of problems and problem solving methods

According to the assembly hypothesis, model-based approaches assume that any expert behaviour can be achieved by

- selecting, adapting, composing or creating a problem solving method, which is a generic conceptual model,

- choosing a shell implementing the method or implementing it by hand,

- adding the application-specific knowledge

They all try to identify various problem solving methods and to classify them. The major distinction is between classification and construction methods. *Classification methods* choose among of a set of pre-enumerated solutions the one best explaining a given set of symptoms. Their subdivision is getting consolidated (see figure 4). In contrast, the classification of *construction methods* like propose and revise, skeletal plan refinement or least commitment which construct a solution that satisfies a given specification, is under heavy debates.

The classification of problem solving methods describes what can be offered to a knowledge engineer from the implementation side, since for each method there are either shells or reference implementations.

In a different hierarchy one tries to catch types of problems as they are encountered by knowledge engineers in real-life. They are characterized in terms of the problem description and the desired solution, but do not refer to any problem solving method. Just like the methods, the problems are subdivided into analytic and synthetic ones. An *analytic problem* is given by a set of symptoms that are to be explained by one or more diagnoses. A *synthetic one* is given by a set of components, construction rules and a set of requirements to be satisfied by the wanted composite object. Again, the analytic class comprising e.g. classification, diagnosis, assessment, monitoring and prediction

```
classification
    simple classification
        flow charts
        decision tables
    heuristic classification
        forward chaining
        backward chaining
        establish-refine
        hypothesize-and-test
    causal classification
        with fault models
        with functional models
    statistical classification
        Bayes'theorem
    case-based classification
        data-base queries
        similarity-based classification
```

Figure 4: Problem solving methods for classification (from [Puppe 90]

```
system analysis
    identification
        classification
        diagnosis
        assessment
    monitoring
    prediction
```

Figure 5: Analytic problem classes (condensed from [Breuker, Wielinga 89])

(see figure 5) is far better understood than the synthetic one which (at least) contains design, configuration, assignment, scheduling and planning problems.

The association between problems and problem solving methods is not unique. For example, if only a small number of constructions exists, then a construction problem might be solved using a classification method that selects the construction best matching the requirements. Moreover, several methods might be combined to solve a problem. Methods applicable to the same class of problems usually differ in the way they expect the problem to be represented and in the additional knowledge they need.

Beside constructing the two classifications, it is an ongoing effort in model-based knowledge engineering to relate them to one another and to find criteria for the most suitable method (combination) for a given problem. The result of this work will allow us to assess the assembly hypothesis.

4   A universal modelling approach

In this section, I will sketch the KADS knowledge acquisition methodology. It will serve as a reference for comparing three prominent shell-based modelling approaches in the next section.

In the KADS esprit project, a vocabulary and a scheme for modelling expertise was developed which has been called *conceptual model* or *knowledge model* [Wielinga, Breuker 86], [Breuker, Wielinga 89]. As shown in figure 6, it distinguishes four kinds of knowledge on four different layers. It is assumed that all application-specific knowledge can be factored out and can be described statically in terms of concepts and relations on the first layer, the *domain layer*. The next layer, the *inference layer*, describes the problem solving competence without any control aspects in terms of metaclasses and knowlege sources. Knowledge sources describe basic inference steps by referring to domain relations. Knowledge sources operate on metaclasses which abstract from the concepts of the domain layer and define the roles they may play during problem solving. Control is specified on the *task layer* in terms of goal-directed tasks which may be composed of subtasks and eventually of knowledge sources. The fourth, *strategic layer* is supposed to monitor and, in case, repair the execution at the task layer. However, this layer is not very elaborated and has hardly been used so far. [1]

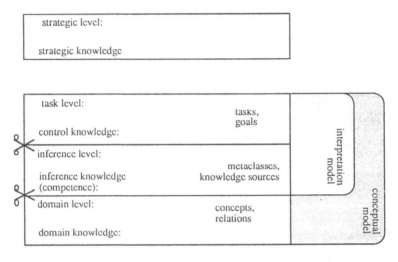

Figure 6: The four layers of KADS' conceptual models

As an example, consider an office-room allocation problem [Karbach, Linster, Voss 89]. Employees are to be assigned to office rooms according to requirements concerning communication, infrastructure, personal features and professional status. The notions of employee, room and requirement are introduced at the domain layer. On the inference layer, employees are abstracted to components and rooms to locations. The inference layer provides knowledge sources for selecting the next components (employees) to be placed, finding the respective requirements, instantiating them to constraints on the components' locations (employees' rooms), and assigning them locations (rooms) while satisfying the constraints. The task layer specifies an initialization step and a selection-satisfaction loop. Figure 7 shows the inference layer with metaclasses being represented by boxes and knowledge sources by ovals.

As indicated in figure 6, the layers can in principle be separated and, ideally, be recomposed. Splitting off the domain layer yields a so-called *interpretation model*, *generic model* or problem solving method which consists of inference and task layers. The task layer, too, may be exchanged. Interpretation models may be composed layerwise and applied to different domain layers corresponding to different applications. For instance, we could use the interpreation model of our office-room assignment system in order to assign hotel rooms to guests, or beds in intensive care stations to patients.

The KADS group started to build a library of interpretation models in order to support a top-down approach to knowledge engineering. As shown in figure 8, the first elicitation phase should aim at substantiating the choice between the various interpretation models. They are subsequently

---

[1]In the ESPRIT Basic Research project REFLECT (project no. 3178), we will probably substitute the strategic layer by a complete metasystem that operates on top of the (object) system modelled by the domain, inference and task layers. The metasystem will in turn be described in terms of domain, inference and task layers [Bartsch-Spoerl, Reinders 90].

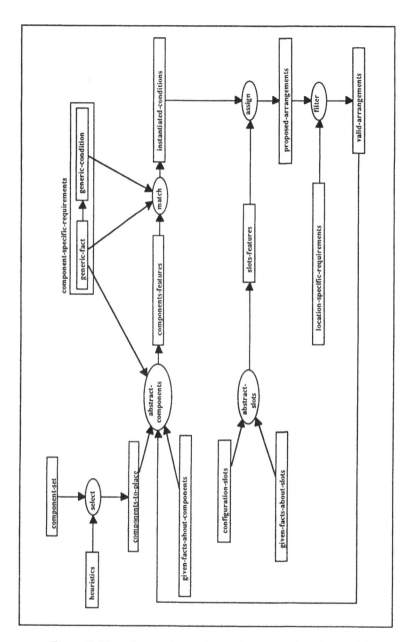

Figure 7: The inference layer of the office room allocation model

composed, adapted and extended by the domain layer. Vice versa, given a self-made conceptual model, the domain layer is removed, the basic interpretation models are disentangled and added to the library.

All constructs in the conceptual model being informal, KADS conceptual models are not executable. According to the KADS methodology, these models serve as input to a separate design phase, where the architecture of the envisaged system is determined and which is subsequently refined and implemented [Schreiber et al. 87].

5 Comparison with some shell-based modelling approaches

I will now briefly compare KADS with three other prominent modelling approaches to knowledge acquisition. A more detailed comparison can be found in [Karbach, Linster, Voss 90]. While KADS is universal, allowing any problems to be modelled, the other three are all shell-based, providing special purpose shells or generating special purpose systems.

In the so-called *role-limiting method approach*, Eshelman, Klinker, Marcus and McDermott have developed several shell/editor combinations [Marcus 88]. The editors are knowledge acquisition tools which provide a conceptual view of the problem solving method and the domain structure realized in the shells. These views are called *role-limiting methods*, as they prescribe the domain structure, i.e. the slots or roles to be filled in. The editors support adding the application-specific contents knowledge. As shown in figure 9, the role-limiting methods comprise, but do not distinguish, KADS' strategy, task and inference layers, as well as the structure of the domain layer. In contrast to KADS, the structure of the domain layer is clearly separated from its contents, i.e. the concepts from their instances and the relations from their tuples. The problem solving method and the domain structure is built-in and cannot be modified. Adding the contents knowledge via the editors yields a knowledge-based system that applies the built-in problem solving method to the specific domain.

The *generic task approach* introduces so-called *generic tasks* to describe problem solving methods together with their input, ouput and knowledge representation. Beside some very special purpose generic tasks, Brown, Bylander and Chandrasekaran developed the languages CSRL and DSPL for applying generic tasks for classification respectively design problems to particular applications [Chandrasekaran 88]. As shown in figure 10, a generic task has the same scope as the shell/editor combinations of the role-limiting method approach. It covers the upper three KADS layers and the structure of the domain layer without explicitly distinguishing them. In contrast to the role-limiting approach, there are no tools supporting the application of generic tasks to a particular domain.

Musen is advocating a two-stage approach [Musen 89] which we have called the *method-to-task-approach* in [Karbach, Linster, Voss 90]. Together with Combs, Fagan and Shortliffe he developed a tool called PROTEGE which provides a conceptual view on the skeletal plan refinement method. Musen calls this view a *method-based conceptual model*. In contrast to the role-limiting method and the generic task approaches, the method-based model covers only the problem solving method or, in KADS terms, the upper layers (c.f. figure 11). PROTEGE supports the knowledge engineer in adding the domain structure which it uses to produce yet another tool. For example, OPAL is the result of applying PROTEGE to the domain of cancer treatment protocols and HTN to treatment plans for hypertension. These tools provide a conceptual view called *task-based conceptual model* which again has the same scope as the role-limiting methods or the generic tasks. The latter tools are intended to be used by the experts to in fill in the domain-specific content knowledge yielding a knowledge-based system, e.g. P-ONCOCYN from OPAL.

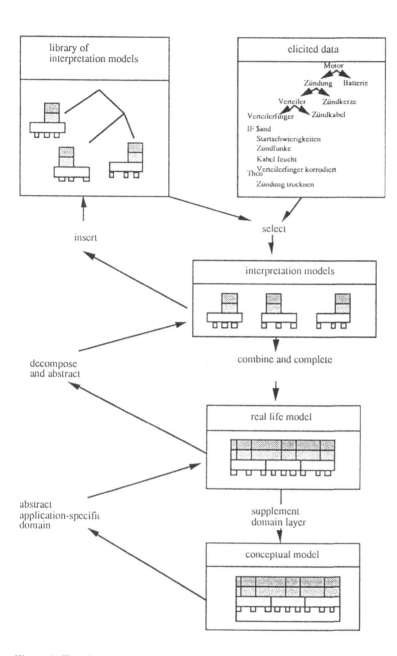

Figure 8: Top down approach to model-based knowledge engineering using a library of interpretation models and bottom-up construction of the library

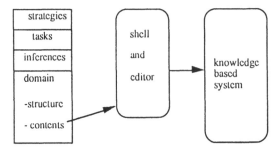

Figure 9: KADS and the role-limiting-method approach approach

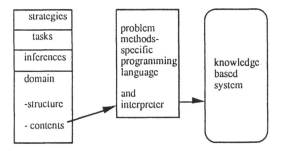

Figure 10: KADS and the generic task approach

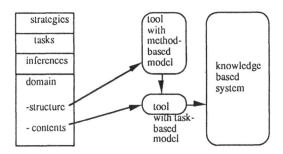

Figure 11: KADS and the method-to-task approach

6 Knowledge engineering as a kind of software engineering

In sections 1 and 2, I have explained knowledge engineering in terms of its two main traditions, rapid prototyping and conceptual modelling. I introduced knowledge acquisition as consisting of knowledge elicitation, structuring and operationalization. I mentioned that knowledge acquisition has been used as synonym to, or subphase or subtask of knowledge engineering, and that the same confusion is encountered between the terms knowledge acquisition and knowledge elicitation. Finally I promised to give my (current) opinion in this section.

When Birgit Ueberreiter saw yet another subtask hierarchy agreed upon in the GI special interest group Knowledge Engineering, she suggested instead to view knowledge engineering as a kind of software engineering. This helps to introduce knowledge engineering to people familiar with conventional software engineering. Moreover, many notions, methodologies, methods, techniques etc. from software engineering might be inherited, if we can get the connection straight. So let me try.

"*Knowledge Engineering denotes both, a discipline and an activity. The subject of the former and the purpose of the latter is the operationalization of expertise on the computer.*" With these words *I set out to write this paper. Now I continue:* "Software engineering, too, is a discipline and an activity which deal with operationalizing an arbitrary behaviour on the machine. Thus, we might view knowledge engineering as a kind of software engineering. But then, knowledge engineering should proceed along the same phases and produce the same kind of documents.

In order to check the latter condition, let us see how we can use knowledge engineering documents (prototype, conceptual model, operatioal conceptual model or a KADS design model) instead of conventional software engineering documents (requirements definition, problem analysis or functional specification, design specification or implementation). As shown in figure 12, a satisfactory prototype may well be sold as the final product, i.e. as the implementation. But it may also be used as a functional specification which is to be implemented according to very different criteria, such as efficiency or programming language restrictions. A conceptual model is intended as a problem analysis or functional specification. It is to be followed by a design specification and implementation, but instead of a conventional software design one could use a design model as suggested by the KADS group. Operational models, too, may be used as functional specifications, but equally as complete implementations.

	requirements definition	problem analysis	design	implementation
prototype		X		X
conceptual model		X		
design model			X	
operational model		X		X

Figure 12: Knowledge engineering documents as software engineering documents

Thus, knowledge engineering documents cannot be uniquely matched with software engineering documents (see also [Tank 90]). Consequently, the phases leading to these documents can neither be matched uniquely. Moreover, knowledge engineering is notoriously cyclic, which counteracts a clean temporal separation of phases. Therefore, I still refrain from separating knowledge engineering into such phases. However, I would like to adopt the software engineering phases (requirements definition, problem analysis, design, implementation, integration/test, maintenance) as knowledge engineering subtasks, thus hopefully inheriting some of the software engineering results. Since the association between knowledge engineering and software engineering documents is not fixed in advance, one of the first jobs of the requirements definition task will be to determine this association for the project.

How does the notion of knowledge acquisition fit into this view of knowledge engineering? Obviously it cannot be a subphase since I have just argued against explicit phases.[2] Neither can we define it as a subtask, because I have just voted for adopting the software engineering subtasks. Therefore I propose to view knowledge acquisition as (the discipline providing) (or as the activity of using) a workbench of methodologies, methods, techniques and tools that are special to knowledge engineering in contrast to software engineering. Figure 13 shows, how the knowledge acquisition subtasks can be used in various software engineering subtasks. Elicitation will be necessary not only in the very beginning, but also during structuring and operationalization, just whenever there is missing knowledge. Furthermore, elicitation and to some extent structuring and operationalization will again be required for maintenance. That means, the same techniques and tools that helped us build a system should again help us in keeping it up-to-date.

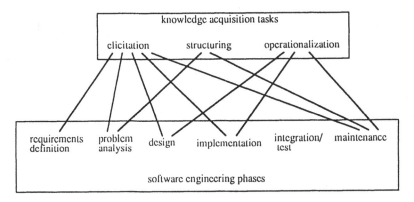

Figure 13: Association between knowledge acquisition tasks and software engineering phases / tasks (from [Ueberreiter 90])

7 Conclusion

This paper began by distinguishing two traditions in knowledge engineering: rapid prototyping and conceptual modelling. Both were further refined into universal and shell-based approaches, the former allowing to describe any problem solving behaviours while the latter are dedicated to a class of problems being handled by a shell. As a synthesis of the two traditions, an approach called model-based prototyping" was favoured that combines the advantages of both. It provides a conceptual description which tries to avoid the implementation details involved in building a rapid prototype, but in contrast to the (universal) model-based approaches this description is operational, thus enabling rapid testing and experimenting.

---

[2]I see no point in defining it as synonymous to knowledge engineering.

Later, knowledge engineering was discussed as a kind of software engineering. It turned out that conventional software engineering phases are not clearly distinguished and that the documents produced cannot be uniquely interpreted as functional specification, design or implementation. However, the software engineering phases might well be adopted as knowledge enigineering subtasks thus avoiding a strict temporal ordering.

Consequently, knowledge acquisition could neither be viewed as a subphase, nor as a subtask of knowledge engineering. Instead, it might be considered as a workbench that supports all aspects of knowledge engineering in contrast to software engineering. Knowledge acquisition was devided into elicitation, structuring and operationalizing subtasks. The knowledge acquisition bottleneck was described and two hypotheses were given as explanations. For the rapid prototyping approach, the mapping hypothesis claims that elicitation techniques must be improved, while for the modeling approaches, the assembly hypothesis states that we must look for the right set of elementary generic problem solving methods (or models). A comparison of elicitation techniques with the types of expertise covered outrules the mapping hypothesis, while the assembly hypothesis is still to be evaluated by the model-based approaches.

According to the assembly hypothesis, all model-based approaches assume that a real-life problem can be solved by a combination of elementary generic problem solving methods. Correspondingly, they try to classify real-life problems as well as problem solving methods and to develop criteria for finding the right problem solving methods for a given problem.

The state of the art in model-based knowledge engineering was presented by four prominent approaches. KADS as the only universal one and the role-limiting method, the generic task and the method-to-task approaches as three shell-based ones.

How will things go on? Will the model-based tradition split into a universal and a shell-based one? Or will the two come together? Is there any chance or are opinions unreconcilable? I am optimistic. At EKAW (European Knowledge Acquisition Workshop) 1990 and at the 5th Knowledge Acquisition for Knowledge-based Systems in Banff 1990 a comparison of the four approaches [Karbach, Linster, Voss 90] which I wrote together with my colleagues Werner Karbach and Marc Linster will be discussed by the community, and I hope this will lead to a common terminology and a comparative evaluation of the four approaches. So let me conclude with the last paragraph from that paper:

"A combination of the four approaches looks promising: the flexibility of KADS to define models, the explicit representation of the operational modelling constructs as it is done in the generic task approach, the role-driven knowledge acquisition capabilities of the role- limiting method approach combined with editors that separate the task of the knowledge engineer from the responsibility of the domain expert, as it is done in the method-to-task approach."

8   Relation to work on databases

Why did I give a talk on model-based knowledge acquisition on a joint database / knowledge representation workshop? Do data-base engineers face a data acquisition bottleneck? Can they profit from knowledge engineering or vice versa? I will use the opportunity of the last section to give a few replies to these questions.

My message:
Beside the egoistic reason of learning about current research on data bases and AI, I wanted to convey a message to the knowledge representation community. Since a long time, I have felt disappointed from talks about knowledge representation. They seem to center around the representation of static, factual knowledge and to ignore the representation of the problem solving method, the

knowledge how you do it given all the facts (c.f. the KL-ONE tradition [Nebel 90]). By the way, this problem is also solved very unsatisfactorily in all shell-based modelling approaches. Only the KADS group has made a constructive proposition, namely to differentiate strategic reasoning, control, and pure problem solving competence or inference knowledge. I hope, the important role problem solving methods currently play in knowledge engineering has become clear from this article so as to be recognized in the knowledge representation community.

Data bases as domain layers?

I have always thought that the domain layer of universal conceptual models might possibly reside in a data base, in particular with domain knowledge increasing or when the data base has existed before the knowledge-based application. The workshop has confirmed this idea. The domain layer with its concepts and relations might be described in terms of an entity-relationship model. However, there is a minor problem with intensionally defined knowledge. Maybe, this can only be represented in a deductive data-base? Or can it be expressed by integrity constraints? Anyhow, the KADS idea of separating the domain layer from the problem solving method nicely fits in. Just like several application programs may operate on a common data base, different problem solving methods may share common (parts in their) domain layers.

Inference and task structures to solve the transfer problem?

Nelson Mattos mentioned the problem of how to cluster the knowledge residing in a data base in order to optimize transfer between the data base and a knowledge-based allication [Mattos 90]. If there is a KADS model for the latter, the required information may be extracted from the task and inference layers. The tasks cluster connected knowledge sources and the inference layer describes the domain knowledge transferred or used by them. Thus, it would be a good idea to load all the concepts and relations addressed by knowledge sources executed in a common task.

No data acquisition bottleneck?

From the workshop I got the impression that database engineers have not yet been confronted with a data acquisition bottleneck comparable to the knowledge engineers' knowledge acquisition bottleneck. Is the information to be put into a database always obvious, lying around waiting to be picked up? Is there no problem of contradicting experts or documents? And is the representation of the information in the database only a question of optimization? Mayr's text analysis [Thoma 90] was the only "elicitation technique for databasesmentioned during the workshop; by the way,I wonder who will type in all the texts to be analysed. Anyhow, if database engineers ever will encounter a data acquisition bottleneck, the elicitation techniques developed for knowledge engineering might be helpful.

No redundancy problem in knowledge engineering?

Helmut Thoma explained that database engineers have problems with keeping the various databases in a big company free of redundance [Thoma 90]. This problem of finding an existing relation that already contains the information wanted arises when there are huge amounts of data and different persons developing and maintaining the databases. Such problems have not yet been dealt with in knowledge engineering. The only project known to me that may have been confronted with them is Lenat's CYC project [Lenat 88]. It aims at representing the knowledge of a complete encyclopedia in a knowledge base. Since 1984, more than half a million entries have been built (and reorganized), and an increase in one order of magnitude is expected by mid-1990, and one more by the end of 1994.

Thus, while awaiting the data acquisition bottleneck and huge knowledge bases, it was a good idea to get acquainted on the workshop. A follow-up workshop would present a good opportunity to discuss a question left open on the panel: What is the difference between data, information and knowledge?

9 Acknowledgements

I am very grateful to the comments of Hans Voss and Werner Karbach, to Werner in particular for supplying all the literature and doing the Latex make-up. Without him, this article would not have been completed in time.

References

[Breuker, Wielinga 89] Breuker, J.; Wielinga, B.: *Models of Expertise in Knowledge Acquisition.* In: Guida, G., Tasso, C. (eds.): Topics in Expert System Design, Amsterdam: North-Holland, 1989, pp. 265–295.

[Chandrasekaran 86] Chandrasekaran, B.: *Generic Tasks in Knowledge-Based Reasoning: High-Level Building Blocks for Expert System Design.* IEEE Expert, Fall 1986.

[Chandrasekaran 88] Chandrasekaran, B.: *Generic Tasks as building blocks for knowledge based systems: the diagnosis and routine design examples.* The Knowledge Engineering Review, October 1988, pp. 183–210.

[Clancey 85] Clancey, W.: *Heuristic Classification.* Artificial Intelligence 27(1985), pp. 289-350.

[Cunis, Guenter, Syska 87] Cunis, R.; Guenter, A.; Syska, I.: *PLAKON: Ein Uebergreifendes Konzept zur Wissensrepraesentation und Problemloesung bei Planungs- und Konfigurationsaufgaben.* In: Proceedings Expertensysteme 87 (Bericht des German Chapter of the ACM 28), Stuttgart: Teubner Verlag, 1987.

[Hayes-Roth, Waterman, Lenat 83] Hayes-Roth, F.; Waterman, D.; Lenat, D.: *Building Expert Systems.* London: Addison Wesley Publishing Company, 1983.

[Johnson 88] Johnson, N.E.: *Mediating representations in knowledge elicitation.* In: Knolwedge Elicitation: Principles, Techniques and Applications (ed. D. Diaper), Ellis Horwood Ltd. Publishers, Chichester, 1988, pp.177-194.

[Karbach, Tong, Voss 88] Karbach, W.; Tong, X.; Voss, A.: *Fillig in the knowledge acquisition gap: via KADSmodels of expertise to ZDEST-2'expert systems.* In: Proceedings of EKAW88 (eds. Boose, Garnes, Linster), GMD-Studie, St. Augustin, 1988, pp. 31-1 - 31-17.

[Karbach, Linster, Voss 89] Karbach, W.; Linster, M.; Voss, A.: *OFFICE-Plan: Tackling the Synthesis Frontier.* In: Proceedings of GWAI89, Berin: Springer Verlag, 1989, pp. 379–387.

[Karbach, Linster 90] Karbach, W.; Linster, M.: *Wissensakquisition fuer Expertensysteme — Techniken, Modelle, Softwarewerkzeuge—.* Muenchen: Hanser-Verlag, 1990.

[Karbach, Linster, Voss 90] Karbach, W.; Linster, M.; Voss, A.: *Model-based Approaches — One Label one Idea?.* European Workshop on Knowledge Acquisition 1990, Amsterdam.

[Lenat 88] Lenat, D.; Guha, R.V.: *The world according to CYC.* MCC technical report no ACA-AI-300-88.

[Marcus 88] Marcus, S. (ed.): *Automating Knowledge Acquisition for Expert Systems.* Boston: Kluwer Academic Publishers, 1988, pp. 81-123.

[Mattos 90] Mattos, N.: *An approach to DBS-based knowledge management..* In: This volume.

[McDermott 88] McDermott, J. : *Preliminary Steps Towards a Taxonomy of Problem-Solving Methods.* In: Marcus , S. (ed.): Automating Knowledge Acquisition for Expert Systems, Boston: Kluwer Academic Publishers, 1988, pp. 225–255.

[Musen 89] Musen, M. A.: *Building and extending models.* In: Machine Learning, Vol.4, 1989, pp. 347-375.

[Nebel 90] Nebel, B.: *Wissensrepraesentation: Probleme und Loesungsansaetze.* In: this volume

[Puppe 86] Puppe, F.: *Assoziatives diagnostisches Problemloesen mit dem Expertensystem-Shell MED-2.* Dissertation, Universitaet Kaiserslautern, 1986.

[Puppe 90] Puppe, F.: *Problemloesungsstrategien in Expertensystemen.* Univ. Karlsruhe, 1990.

[Bartsch-Spoerl, Reinders 90] Bartsch-Spoerl, B.; Reinders, M.: *A tentative framework for knowledge level reflection.* deliverable of task I.3, to appear.

[Schreiber et al. 87] Schreiber, G.; Bredeweg, B.; Davoodi, M.; Wielinga, B.: *KADS: B2-Design.* Esprit Project P1098, VF Memo 97, November 1987.

[Tank 90] Tank, W.: *Ueber das Selbstverstaendnis der Disziplin des Knowledge Engineering.* KI-2-90, Oldenbourg Verlag, Mnchen, Juni 1990.

[Thoma 90] Thoma, H.: *?????.* In: This volume

[Ueberreiter 90] Ueberreiter, B.: *Modelbasierte Wissensakquisition - Anforderungen aus der industriellen Praxis..* In: Tagungsband des zweiten Workshops der GI-Fachgruppe Knowledge Engineering, Daimler-Benz Arbeitsberichte, Berlin April 1990 (to appear).

[Voss et al. 90] Voss, A.; Karbach, W.; Drouven, U.; Lorek, D.; Schukey, R.: *Operationalization of a synthetic problem.* deliverable of task 1.2.1, REFLECT-report RFL/GMD/I.2.1, GMD St. Augustin, April 1990.

[Wielinga, Breuker 86] Wielinga, B.; Breuker, J.: *Models of expertise.* Proceedings of ECAI86, Brighton, 1986.

Practical Experiences - a Panel Session

Helmut Thoma
CIBA-GEIGY Ltd.
Basle (Switzerland)

At the well attended panel session - held in the evening after the buffet - the participants of the workshop discussed the difficulties which practice has still to overcome in integrating classical information systems and expert systems. With the moderation of the chairman, Helmut Thoma of CIBA-GEIGY Ltd., Basle, Hans-Peter Hoidn of the Institut für Automation AG, Zurich, Nelson Mattos of the University of Kaiserslautern and Wolfgang Sager of the COLLOGIA Unternehmensberatung GmbH, Cologne presented some important topics of integration and had a useful discussion with the audience.

H.-P. Hoidn reported on the implementation of a knowledgebase-database-coupling in a large Swiss bank. A planning system for batch processing in the computer center of the bank was implemented using KEE, COMMON-Windows, ORACLE and KEEconnection for the linkage of object-oriented systems and database systems. Central to presentation of N. M. Mattos were the experiences with performance measures of coupled (relational) databases and expert systems. Such measurements and analysis can seldom be realized in practice beyond research institutes. In businesses knowledge-based systems should be integrated as an implementation of innovative technology in existing environments on a technical, professional, administrative and user level. These and related viewpoints were discussed by W. Sager in connection with his experience with Artificial Intelligence technology in conventional business surroundings.

Loosly-coupled AI systems and conventional systems have problems with integrated development tools. In addition there is in practice unfortunately a lack of project leaders and project employees with a knowledge of both technologies. Also the cost of updating knowledge in the operational phase of knowledge based systems is frequently underestimated.

The design of the object classes and relations should originate from a common data model (e.g. a common Entity-Relationship Model). Special attention should also be given to the integrity of the data, to the object system and to the design of the transactions.

Expert systems most frequently retrieve isolated, scattered data; this is inefficiently handled by today's database systems. Effective cooperation between the two types requires better adjustment to the locality of the data. The storage structure of the database system should be adjusted to the way expert systems work and expert system calls should be better matched to the set-oriented interfaces of relational database systems. To sum up, an optimal expert system / database system coupling requires a better coordination of tasks and an appropriate defined interface. These facts have been experimentally demonstrated.

Below, the reader will find more details in statements prepared by the three members of the panel session.

Practical Experiences in Coupling Knowledge Base and Database in a Productive Environment

Hans-Peter Hoidn

IFA, Institut für Automation AG, Zürich

Abstract

Integration of database and of object–oriented environment is crucial for future information systems. This paper presents an example of a today's solution of an object–oriented system coupled with a relational database to obtain a multiuser system. The practical experiences were gained with *MetaPSS* , a planning system for batch processing. They cover modelling and implementation issues. The modelling has to allow a mapping of the object structure onto a database structure. For the implementation we discuss the transfer mechanisms which must support the mapping of objects to data and the transaction design which has to guarantee the data integrity.

1 Introduction

This paper presents some practical experiences from the realisation of *MetaPSS* , summarizes the presention and includes issues of the panel discussion. *MetaPSS* is a knowledge-based system coupled with a relational database system for the planning of the batch processing in the EDP environment of *UBS* (Union Bank of Switzerland). Main issues of the discussion are: The design of the whole system is carried out with an object design which can be mapped onto an entity relationship model. In contrast to other implementations of coupling a knowledge base and a database, *MetaPSS* needs both directions of mapping. Thus maps are needed to transfer objects into tupels in the database and vice versa. The coupling includes the design of transactions: one user session is a long transaction. Locks on the database preserve consistency of the data. Examples and solutions drawn from *MetaPSS* are provided.

The paper is organized as follows: In the following section we present a brief overview of *MetaPSS* , an expert system for production planning of batch processing. In Section 3 the modelling and the design methodology is described. Section 4 contains a discussion of some implementation issues like the transfer mechanisms and Section 5 describes the transaction design.

2 METAPSS

This section gives a brief description of *MetaPSS* , more about *MetaPSS* can be found in [7].

2.1 The application background

MetaPSS is now a delivered expert system for the planning of batch processing in the productive EDP environment of *UBS*. In this planning, Cobol programs (or *tasks*) belonging to the same application are grouped in *runs*. Thus, a *run* consists of several *tasks* and their relations. The *tasks* are performed in a predescribed order which can be presented in a flow diagram. Each *task* references *resources* (files and areas) with a reference mode (read, write), has successors, belongs to a *rollback cycle* etc. By this informations, *MetaPSS* generates the actual job control, which executes Cobol programs at run time.

2.2 Hardware and Software Base

MetaPSS was developped on SUN–Workstations 3/60 and 3/280. Now the productive system runs on the same environment. *MetaPSS* is a *LISP* based system and is implemented with the object system of *KEE* , which is an object–oriented environment providing canonical object features like classes, inheritance, methods and message passing.

The highly graphical interface of *MetaPSS* was implemented with COMMON–Windows (from *IntelliCorp*). The database, running on a dedicated database server (SUN 3/280) is implemented with the *ORACLE* –RDBMS. *KEEconnection* (from *IntelliCorp*) is used for the implementation of coupling knowledge bases and databases.

An unexpected side effect concerned the proper releases of all the software tools used. As in practice it is not obvious that in all cases you have the right versions from different software vendors at hand. SunOS, Lucid LISP, *KEE* , *ORACLE* RDBMS, *SQL*PLUS* and *KEEconnection* must be installed in corresponding versions. We ran SunOS 4.0. *KEE* worked fine. For the *ORACLE* RDBMS however, some additional support was necessary, since the available *ORACLE* –version 5.21.3 was not foreseen to run under SunOS 4.0. Only the installation of *ORACLE* –version 6.0 allows to install the latest version of *KEEconnection* .

2.3 System architecture

MetaPSS contains two types of permanently loaded *knowledge bases*: the abstract knowledge about runs (applicatorial rules and the syntax of *PSS* itself) is stored in *design knowledge bases*, *frame knowledge bases* contain the class structure, i.e. each instance of a concrete run inherits from a class of a *frame knowledge base*.

MetaPSS supports an intelligent and highly graphical editing of runs:

- A *Flow Editor* allows the graphical editing of tasks, task groups and their relations, i.e. adding or removing successor relations. Moreover, run conditions are shown.

- With the *Hierarchy Editor*, the set structure of a run, consisting of task groups (which e.g. are rollback cycles) and tasks can be designed or modified.

- An "intelligent" *Task-Description-Interface* of a task's description allows to work on the job stream specifications. Such a task description consists of all informations needed by the executable job stream. *MetaPSS* contains a knowledge base which stores the description syntax and the related methods. Thus, input can immediately be checked.

MetaPSS is a multiuser expert system. The object world contains the context and the related methods. The information of a specific run is permanently stored in the database, in a working session the information is mapped onto instances in the object system. *MetaPSS* contains the functions to *download* information from the database and to build up the instances; this is done at the beginning of and during a user session. Moreover, *MetaPSS* performs the *upload*, which stores back all modifications and new instances; this is done at the end of a user session. The context, stored in *design* and *frame knowledge bases* and the related methods are all contained in one *LISP* world (which is loaded from a disk save file).

Compared to other knowledge systems (like the *Airbus* development as mentioned at the panel discussion), the classification of the object world in *MetaPSS* leads to a relatively small set of classes, each class containing a large number of instances (even for one run). E.g., all tasks have the same structure (referenced resources, run conditions, etc.); thus a lot of task instances use the same task class structure and methods. The clean separation of abstract knowledge from the actual data mentioned above, allows to store only the values of the instance variables, and this is crucial for the whole coupling design.

Notice that the *KEE* –rule system has not been used up to now since the modelling was done completely with objects containing data and methods.

2.4 Project History

The development of *MetaPSS* was started in 1986 on *Symbolics*–workstations. In april 1988, a *demo prototype* (see [2]) demonstrated requirements of the user interface containing preversions of the flow editor and the hierarchy editor. Next, the *full prototype* was started with a team of 3 to 5 persons and was finished in time at the end of march 1989. The *beta release* of the delivery version with the integrated database was finished at the end of the year 1989. This was in time according to the project plan of april 1988. Now in 1990, two persons are maintaining the system, stabilizing it, improving the performance by tuning the database management system and working up new user requirements. The extensive use of the system will show, whether the overall performance will be acceptable for the users. Presently, (without tuning the database) they are not enthousiastic about it.

3 Modelling

3.1 Approach

As described in the history of the project *MetaPSS* , the development started with the object system to build up the *class structure* in *frame knowledge bases*. A prototyping approach was chosen for this task. For the design of the data base, the class structure obtained in the object system was transformed to an *entity–relationship–model* which uses the concepts of entities, relationships and attributes to represent data. For this part of the work, a classical approach of software engineering was applied.

The development of *one* system needs a consistent design, for *MetaPSS* it has to be valid for the object system *and* the database. The need of an "unifying design" is widely accepted (see i.e. [6]). Such a general design can not be covered by a prototyping approach. Rather proven methods of software engineering should be used. The main focus of prototyping still remains for

TASKS

IDENTIFICATION	..	..
..	..	..
EBLINF+8912	..	..

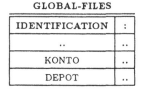

#[Unit:KONTO]	
IDENTIFICATION	
KONTO	
.....	
.....	
REFERENCED-BY	
(#[Unit: EBLINF+8912])	

TASK-RES-REF

TASK-ID	RES-ID	..
..	..	..
EBLINF+8912	KONTO	..
EBLINF+8912	DEPOT	..

#[Unit:DEPOT]	
IDENTIFICATION	
DEPOT	
.....	
.....	
REFERENCED-BY	
(#[Unit: EBLINF+8912])	

GLOBAL-FILES

IDENTIFICATION	:
..	..
KONTO	..
DEPOT	..

Table 1: database design

the design and the "look-and-feel" of the user interfaces. In the project team of *MetaPSS* , the paradigm shift from prototyping to modelling caused no problems. A reason might be, that all team members held a degree in mathematics or physics and were not focussed on AI, especially.

3.2 Logical Mapping of Structure

The correspondance of the structure of the object system to the structure of the database can be described by logical mappings.

KEE [5] provides *classes* and *instances* of classes, and inheritance may occur between a class and a subclass as well as between classes and instances. Thus, classes can be ordered in a directed *class structure*. Instance variable names are called *value slots, method slots* contain procedures.

In the traditional approach of an *entity-relationship-model*, each class corresponds to an *entity set*. A value slot (of a class) corresponds to an *attribute* or, if this slot contains references to other objects, to a *relationship*. n:m–relationships are allowed (for example in Table 1 with REFERENCED-RESOURCES and REFERENCED-BY). In the object system, the relationship A–B is represented by the value B in object A. Allowing redundancy (however, redundancies only exist in the object world and not in the database), A–B may be represented by the value A in object B as well.

In the *relational database*, one object (i.e. one entity) corresponds to one tuple in the so–called *primary table*. A column of the *primary table* corresponds to a slot (an attribute) with cardinality 1. A value slot may have multiple values (in the terminology of Smalltalk a *collection class*). Due to the normalisation, a slot (i.e. an attribute) with multiple values corresponds to a so called *secondary table*. Such a slot of an object may contain a set of values which must be mapped one-to-one to a set of tuples in the corresponding *secondary table*. To map one column value to one slot value, a mapping function is needed. A 1:n–relationship between two objects is represented by one column value in one primary table; a n:m–relationship by one tuple in a so called *relationship table* (see Table 1).

3.3 Practical Aspects

3.3.1 Identification of Objects

In a single–user object–oriented environment, naming of objects is not a problem and could be done by a name generator using an internal symbol table. A logical database design, however, needs key attributes which are unique not only in one working session but in the whole data world. Thus, objects in the object world of a coupled system need to be identified in an unique manner. In *MetaPSS* , this is solved by a special attribute, called IDENTIFICATION (for example EBLINF+8912). Uniqueness of the keys may be implemented with unique indexes.

3.3.2 Integration of two different "worlds"

An advantage of object orientation is, that objects are modules containing both data and behavior, thus objects represent some knowledge. This approach allows to design and implement complex systems and results in a more compact design of a software structure. Without the use of object–oriented programming, *MetaPSS* could not be realised, since the complexity of the system is higher than tools of relational database systems can manage. Moreover, the object system itself can have a higher degree of integrity than a conventional software system. This has been experienced when one object is usually mapped on a bunch of tuples in different database tables. Thus, the advantages of encapsulation which preserves some integrity is lost.

3.3.3 AI and DB

"AI and DB differ completely" [1]. This is reflected by the structure of knowledge bases and relational databases. Knowledge bases have a complex data structure, objects may contain objects etc. Moreover, the objects' behavior is determined by their methods. Reasoning in knowledge bases is highly sophisticated and reflects all the possibilities of such a system. Relational databases however, have a simple structure of tupels, each tupel containing strings and/or numbers. They are designed to have fast access to data in huge amounts of data. Reasoning in databases is done by query along some attribute values.

The integration of knowledge bases and relational databases requires a bridge between the two different "worlds". This implies, that complex data structure must be mapped onto simple ones and vice versa, for which *KEEconnection* provides a good tool base. Moreover, one object is mapped on a bunch of tuples in the database, this task is supported sufficiently by *KEEconnection* too. Obviously, these transformations consume much CPU–power.

```
#[Unit: EB+8912.03]
identification
EB+8912.03
....
#[Unit: IN]
....
HTO
....
((#[Unit: OPTIONS].T)
(#[Unit: NOUGHT]))
```

TASK-E-EXPRESSIONS

IDENT	ENV-SEL	KEY-TOKEN	PARAMETER
...	...	...	...
EB+8912.03	HTO	IN	/OPTIONS./T;NOUGTH
...	...	...	...

Table 2: KB-DB translations

4 Implementation

4.1 The mapping and translation module

Mainly this module supports the translation between equivalent representations of the same entity. For each entity set has to be specified:

1. the way columns (of the primary table or of the secondary tables) map onto the slots of the corresponding *frame knowledge base* class (*logical mapping design*)

2. the function providing the translation between a set of database fields (a row in the database) and *KEE* objects as well as its inverse (*implementation*)

The logical mapping design is entirely specified by the development interface of *KEEconnection* ; this powerful graphical tool (the *Mapping editor*) provides an adequate way to define the structure of the mapping as well as of the joins between primary and secondary tables. This structure is then stored in a *mapping knowledge base*, which contains all the information needed to map the tables onto the *frame knowledge base* classes and which is loaded together with the *MetaPSS* knowledge bases.

4.2 Practical Aspects

The *mapping* is well supported by the tool *KEEconnection* , which provides some elementary translation functions. For *MetaPSS* , additional *LISP* functions for the transformation of database strings to complex *LISP* structures were written, which now extend the translation module.

All slots of all objects contain their history: a facet named *Retrievedp* contains the state of the value: either it is untouched (no download performed), downloaded or modified. Thus a close coupling of the knowledge base and the database can be realised: if an untouched slot is used, the download is performed first. If an upload (update to the database) is started, all unmodified slots remain untreated, only the modified and new slots are uploaded.

5 Transaction Design

To keep the data consistent in the multiuser environment of *MetaPSS* , one user session is designed as a long transaction. The transaction design of *MetaPSS* covers two main aspects:

1. Users of *MetaPSS* act in a multiuser environment. Thus, the transaction design has to guarantee, that at the same time, not more than one user can edit the same run and cannot access the same data for modification.

2. The modifications of a user session must be stored back to the database in such a way, that at any time, the database remains consistent or can be rolled back to a consistent state.

These two aspects are covered in the following description of a user session. We refer to the following definitions of *"coupling"*:

1. *Loose coupling* means, that the object system and the database are independent. At the beginning of a user session, data are transferred from the database to the knowledge base, e.g. by a file transfer.

2. *Tight coupling* uses a database as a backend of the knowledge based system.

5.1 Locking

A run is the main logical entity for one user session (from an applicational point of view). Thus, the parallel editing of the same run makes no sense and, therefore, is not allowed. Since it is possible to edit some resources (files or areas) which are referenced by more than one run, the parallel editing of a run and of a strategic resource referenced by this run is not allowed either. These features are implemented with database locks (row locking). Thus, the transaction design implies, that the whole data are divided in disjunct subsets (see also [4] and its approach of *disjunct objects* for locking strategies in object–oriented databases).

The first input at the beginning of a usual user session for editing a run is the *run identification*. Then, if possible, a row lock in exclusive mode is done for the tupel representing this run, otherwise the input is rejected as there would be another concurrent edit session. The modification of a resource needs locking of all rows corresponding to runs which have chosen a reference to the *resource*. Notice, that row locking is not provided by all RDBMS systems.

Our approach, to base the multiuser environment on a separation of the data into disjunct subsets, is closely related to the applicational point of view. For *MetaPSS* , this approach is very efficient, but in general it cannot be applied. Thus, it will need a great effort to obtain efficient locking strategies which are general but not application dependent.

5.2 Download

The *download* transfers data from the database to the object system, from the multi user environment to the local working environment. This reflects the aspects of a loosely coupled system. *KEEconnection* provides a *SQL* -like query language which is fully integrated in *KEE* ; up to 1'000 instances can be downloaded in 2-3 minutes. *KEEconnection* allows an efficient and packed download of all needed data (which then passes through the translation module) and it is integrated in *KEE* 's rule-system, allowing reasoning about data in the database. Here are two examples drawn from the productive code:

- (DOWNLOAD '(AND (?TG IS IN (TASK-GROUPS))
  ((THE IDENTIFICATION OF ?TG) IS LIKE "WH1%"))))
  (downloads all the taskgroups of the run WH1)

- (DOWNLOAD '(AND (?T IS IN (TASKS)
  ((THE IDENTIFICATION OF ?T)
  IS CONTAINED IN '("WH1START" "WH1POST")))))
  (downloads all tasks whose identification is "WH1START" or "WH1POST")

At the beginning of a user session, according to the run to be edited, *MetaPSS* downloads the data most likely to be used such as tasks, task groups, resources etc. This initial download takes now 10-12 minutes.

5.3   User Session

The described download at the beginning of a user session is done for the data which are needed in nearly all editing sessions of a run. This initial download reflects the "loose coupling" of the knowledge base to the database. Moreover, there are data of a run, which are fetched only when they are needed. This reflects the "close coupling" of the knowledge base to the database.

In *KEE* combined with *KEEconnection* , the object system keeps track of downloaded or updated slot values using the build-in facet *RetrievedP* (see section 4.2). Thus, data are downloaded only once and all modified slots are known. This feature is very efficient and well integrated. It may be mentioned, that each object system realising the "close coupling" needs such an built–in feature to keep track of each slot's history.

During the user session, no modifications are stored back to the knowledge base (with the exception of *resource identifications*, which are stored back immediately to prevent that one identification can be chosen twice). Thus, only the (temporary) instance knowledge contains the work of a current edit session. Moreover, *MetaPSS* contains local rollback features, e.g. such that a rollback is done when quitting a window.

5.4   Upload

At the end of a session, the *upload* stores back all modifications and new instances: insert, delete and update statements have to be applied to the database . Due to the facet *RetrievedP*, *KEEconnection* performs an upload only for new or modified slots. Joins are handled correctly: for multiple values a test against the contents of the database checks whether an insert or delete, nothing or both have to be performed. A new relationship between objects should for instance result in an insertion of a whole new row in the corresponding secondary table (such a behavior is defined through the graphical development interface).

KEEconnection is primarily download-oriented. For uploads only a small amount of some modified values is foreseen. Therefore we obtained a serious performance problem while uploading a whole new run or even some new parts of a run: the upload sometimes lasted more than one hour. This problem arises, because *KEEconnection* inserts a new row in the primary table by updating field by field. Bypassing some security mechanisms of *KEE* (which are not necessary for completely new instances) *KEEconnection* was extended to efficiently insert new rows as a whole with one or more *SQL* –Inserts.

In our implementation the complete upload is carried out at the end of the user session. Since two concurrent uploads might produce deadlocks and some unsuitable reactions of the RDBMS might destroy the integrity of the database, the final save of a user session including the upload to the database is performed in single user mode. An exclusive lock of all tables during the upload is set to prevent other users from database access during some seconds (rarely minutes). After the upload is performed successfully (otherwise it is rolled back) the final commit releases not only the lock of the tables but the row lock of the run as well.

References

[1] Brodie, M.L. *Future Intelligent Information Systems: AI and Database Technologies Working Together* in Mylopoulos, J. & Brodie, M.L. (eds.), *Artificial Intelligence & Databases* Morgan Kaufmann Publishers Inc. (1989).

[2] Cupello, J. M. & Mishelevich, D. J., *Managing Prototype Knowledge/Expert System Projects.* Communications of the ACM, *31* 5(1988), 534-541

[3] Goldberg, A. & Robson, D., *Smalltalk 80: The Language and its Implementation* Addison-Wesley, 1983.

[4] Herrmann, U. & Dadam, P. & Küspert, K. & Schlageter G. *Sperren disjunkter, nicht–rekursiver komplexer Objekte mittels objekt- und anfragespezifischer Sperrgraphen* in Härder, T. (Hrsg.) *Tagungsband der GI/SI-Fachtagung "Datenbanksysteme in Büro, Technik und Wissenschaft" Zürich, März 1989*, Informatik–Fachberichte 204, Springer–Verlag, 1989, 98-113

[5] IntelliCorp *KEE Version 3.1* Software Development System User's Manual, 1986.

[6] Risch, R. & Reboh, R. & Hart, P. & Duda, R., *A Functional Approach to Integrating Database and Expert System.* Communications of the ACM, *31* 12(1988), 1424-1437

[7] Schlegel, D. *MetaPSS – A Knowledge System for the Design and Maintenance of an Automated Batch Processing*, Tagungsband der SGAICO–Tagung, Lugano, Mai 1989.

Performance Measurements and Analyses of Coupling Approaches of Database and Expert Systems and Consequences to their Integration

Nelson Mattos

University of Kaiserslautern

Department of Computer Science

P.O. Box 3049, 6750 Kaiserslautern, West Germany

e-mail: mattos@informatik.uni-kl.de

Abstract

This paper presents some important requirements to be fulfilled by coupling/integration approaches of database and expert systems, which resulted from detailed investigations of some concrete prototype systems.

1. Introduction

Early knowledge-based systems (KS) were characterized by large varieties, but small quantities of knowledge. For this reason, they emphasized the richness, flexibility, and the accuracy of the knowledge representation schemes and ignored the efficient access and manipulation of the knowledge base (KB) contents [VCJ84]. Mostly, KS applied very simple search algorithms and delegated all KB management tasks to the programming language interpreter being used. Nevertheless, the inefficiency of knowledge handling was not a critical issue because it was compensated by the small sizes of the KB, which could be kept in virtual memory.

Real world applications require, however, the use of large amounts of knowledge for solving problems. KB contents exceed in these cases virtual memory sizes so that they must be maintained on secondary storage devices being managed (for example) by a database system (DBS). With the use of secondary storage devices, efficiency, particularly time efficiency, by storing, accessing the knowledge and making it available for problem solving became therefore the major problem of concern in this environment. Consequently, the investigation of the behavior of KS running on these environments has turned out to be a prerequisite in identifying the specific requirements to be fulfilled here. It is important to investigate the characteristics of KS accesses to the KB as well as in the necessary structures to map knowledge in great depth in order to define adequate mechanisms to increase performance. Because of this, we have adapted some available KS to work on coupling environments as well as constructed some further prototypes in order to pursue this investigation [Ba87,Ma86,St86,Th87].

Some of our observations were already discussed in [Ma90,HMP87,HMM87], in which we presented some of the experiences gained by the analysis of coupling prototypes as well as by the mapping investigation of knowledge structures on data models. In this paper, we abstract the particular conclusions of each of these concrete investigations, putting together those results which will dictate the specific requirements to be supported by environments for coupling/integrating KS and DBS. Further requirements which are closely related to traditional DBS features will not be discussed but are, however, equally as relevant as the following.

2. Analysis Results

Access characteristics

The accesses made by KS at the DBS interface are mainly to tiny granules. In frame-based KR schemes, for example, they refer to individual attributes rather than to objects as a whole. Sequences of accesses to attributes of the same object are also typical. This occurs because inference engines can only process one "piece of" knowledge at a time so that accesses to very small amounts of information and consequently many DBS calls result. Therefore, it is important to have mechanisms that enable the reduction of the path length when accessing KB contents, allowing the KS to reference KB objects directly. In other words, we propose (for example) the design of a component between KS and DBS aimed at high locality of references to KB contents.

Access types

In general, during a KS consultation accesses to the KB are mostly referred to read and write operations. Insert and delete operations may also occur, but with very low frequency since these operations are associated with some kind of structural change of the KB. In other words, during a consultation, modifications in structures of the KB (e.g., changing object types) are very seldom. Clearly, during the KB construction process, this situation is just the inverse. This strong stability should be exploited by DBS when choosing storage structures for KB. Since KB construction is an iterative and incremental process and consequently does not require a performance as high as KS processing, DBS should give more priority to the optimization of retrieval operations.

Access frequencies

The frequency of access to object attributes also differ very much. Whereas dynamic attributes are accessed with high frequency, static attributes are accessed with low frequency [Ma90, HMP87]. In order to make use of this KS access characteristic, when storing KB objects, DBS should split them, keeping dynamic and static attributes separate. By exploiting this kind of clustering in the storage structures, the locality of the KS accesses will be surely better supported. Furthermore, when the KS refers to dynamic attributes (what occurs in most cases), the DBS may access only these. As a consequence, I/O operations as well as transfer overhead will be strongly reduced since the dynamic attributes are much smaller than the static ones. In the case of the investigated KS, for example, they just represent about ten percent of the KB contents.

Processing contexts

It has been observed that in each phase of the problem solving process, KS accesses concentrate on the contents of just some objects of the KB. The reason for this is that KS need different parts of the KB to work on different problems. The accessed objects represent, therefore, the knowledge, the so-called processing contexts [HMP87], needed to infer the specific goal of that phase. The existence of such processing contexts depends, therefore, on the problem solving [St82,Pu88] strategy being used. For example, KS exploiting constraint-propagation have contexts corresponding to the several constraints defined in the KB. In the same manner, a problem-reduction strategy determines contexts for processing each partitioned subproblem, and generate-and-test organizes KB contents according to the KS pruning activity.

Although processing contexts are known just after the definition of the problem solving strategy and the construction of the KB, during a KS consultation, the context needed in each phase can be neither static nor determined at the beginning of this consultation. Usually, it is dynamically established by some information that the KS deduces during the preceding phases (i.e., dynamic knowledge) so that it is possible to determine the next context needed only at the end of a phase. For example, in a hypothesize-and-test environment contexts needed in test phases depend directly on the suspicions generated in the foregoing hypothesize phases. This knowledge about the behavior of KS accesses disables therefore a static but enables a dynamic preplanning of the accesses to the KB that should be used by the DBS for optimization purposes. DBS should exploit the existence of such processing contexts building further clusters in the storage structures as well as applying more appropriate access paths. Such measures will then improve the DBS support of the locality of KS accesses, increasing its performance as a whole.

Temporary maintenance of dynamic knowledge

Since the values of the dynamic knowledge are offered during a consultation and will be, of course, different from case to case being analyzed by a KS, they generally lose their significance at the end of a consultation (they might be relevant for history purposes). The static knowledge on the other hand represents the expertise of the human experts and is, for this reason, relevant for each KS consultation. Because of this, the values of the dynamic attributes are to be kept only temporarily by the DBS, whereas the static ones are to be stored permanently.

Multi-user environments

One can imagine that in multi-user environments, the static attributes are accessed concurrently by each user in order to infer the values of the dynamic ones corresponding to the case which the KS is analyzing for each one. Thus, keeping just one copy of the static knowledge would be sufficient since it will be accessed by multiple users. However, the DBS must maintain for the dynamic knowledge as many versions (i.e., some kind of private data) as the number of users working with KB. These versions are then accessed individually since the values of the dynamic attributes of one user have no meaning to the other ones. Thus, in multi-user environments synchronization should be controlled only for static knowledge.

Knowledge structures

Knowledge structures of KS are usually very complex. Often, KB contents are defined as a composition of other KB contents so that DBS must cope with the handling of complex objects. Furthermore, most KS present not only the usual single-valued attribute types. Frequently, attributes are multiple-valued and present data types (lists, texts, and procedure) whose length may vary extremely. In general, it is not possible to fix the length of an attribute, since it may change dynamically during the consultation, and even not its maximal length. Consequently, DBS in these environments should make use of more sophisticated storage structures and access paths as those normally provided by DBS.

Delegation of KB operation

It has been observed that some kind of inferences can be performed more efficiently by DBS. For example, in [Sm84], it is argued that the two principal performance bottlenecks when combining KS and DBS techniques are the inference process in KS and the access to secondary storage devices in DBS. In trying to solve the first bottleneck, it is concluded that query evaluation can provide some but not all inference capabilities. However, since DBS provide these capabilities much more efficiently than KS, the solution to this first bottleneck is to combine both systems in such a way that the delegation of inference steps to the DBS becomes an easy issue. This gives DBS more margin for exploiting optimization as well as parallel processing what seems to be the solution to the second performance bottleneck.

3. Summary

Based on the storage of the entire KB in virtual memory and, therefore, neglecting the existence of access problems, time has not traditionally been a criteria to measure the effectiveness of existing KS. Usually, this measurement was made by only considering the power of the features of the offered knowledge representation scheme and inference engine. In fact, both KR schemes and inference machines ignore the issues of how the knowledge is managed, by relegating this task to the operation system. However, when KB exceed virtual memory sizes, inefficiency in knowledge management becomes apparent and critical.

The solution to this problem is the delegation of knowledge management tasks to an independent system such as a DBS. Approaches combining KS with DBS for this purpose have shown that the working method of KS are quite different than those of traditional DBS applications. For this reason, by the development of architectures combining KS and DBS techniques, some important issues have to be considered in order to achieve satisfactory performance. Most important is the framework to be provided in this environment for the exploitation of the application locality. In other words, it is desirable to have mechanisms aimed at high locality of KB references to

- enable the reduction of the path length, and
- offer fast access to stored objects.

Second, it is important to have mechanisms to allow for splitting static and dynamic knowledge, for specifying processing contexts, and for exploiting them at level of storage structures and access paths. These issues will

- provide means for more efficient KB accesses,
- reduce I/O operations as well as transfer overhead, and
- enable a more effective use of (application) buffers.

Third, it is necessary to allow for an appropriate (i.e., flexible and efficient) representation of knowledge structures at the DBS level and for the delegation of KS inferences. This issues will

- guarantee a more efficient KB manipulation and
- permit an adequate support of knowledge modeling.

Finally, some further features should reflect traditional DBS characteristics in order to

- efficiently cope with multi-user access,
- guarantee the maintenance of the KB semantic integrity,
- allow for KS exhibiting high reliability as well as availability, and
- permit the distribution of the KB.

References

[Ba87]   Bauer, S.: A PROLOG-based Deductive Database System (in German), Under-graduation Final Work, University of Kaiserslautern, Computer Science Department, Kaiserslautern, 1987.

[HMM87]   Härder, T., Mattos, N., Mitschang, B.: Mapping Frames with New Data Models (in German), in: Proc. German Workshop on Artificial Intelligence GWAI'87, Springer Verlag, Geseke, 1987, pp. 396-405.

[HMP87]   Härder, T., Mattos, N.M., Puppe, F.: On Coupling Database and Expert Systems (in German), in: State of the Art, Vol. 1, No. 3, pp. 23-34.

[Ke86]   Kerschberg, L. (editor): Proceedings from the First International Workshop on Expert Database Systems, Kiawah Island, South Carolina, October 1984, Benjamin/Cunnings Publ. Comp., Menlo Park,CA., 1986.

[Ma86]   Mattos, N.M.: Mapping Frames with the MAD model (in German), Research Report No. 164/86, University of Kaiserslautern, Computer Science Department, Kaiserslautern, 1986.

[Ma90]   Mattos, N.M.: An Approach to DBS-based Knowledge Management, in these proceedings.

[Pu88]   Puppe, F.: Introduction in Expert Systems (in German), Springer-Verlag, Berlin, 1988.

[Sm84]   Smith, J.M.: Expert Database Systems: A Database Perspective, in: [Ke86] , pp. 3-15.

[St82]   Stefik, M.J., et al.: The Organization of Expert Systems, A Tutorial, in: Artificial Intelligence, Vol. 18, 1982, pp. 135-173.

[St86]   Stauffer, R.: Database Support Concepts for the Expert System MED1 (in German), Undergraduation Final Work, University of Kaiserslautern, Computer Science Department, Kaiserslautern, 1986.

[Th87]   Thomczyk,C.: Concepts for Coupling Expert and Database Systems - an Analysis Based on the Diagnose Expert System MED1 and the Database System INGRES (in German), Undergraduation Work, University of Kaiserslautern, Computer Science Department, Kaiserslautern, 1987.

[VCJ84]   Vassiliou, Y., Clifford, J., Jarke, M.: Access to Specific Declarative Knowledge by Expert Systems, in: Decision Support Systems, Vol. 1, No. 1, 1984.

Integration of AI Systems in Conventional Environments

Wolfgang Sager

C O L L O G I A
Unternehmensberatung GmbH
Ubierring 11 5000 Köln 1

1 INTRODUCTION

During the last two decades some AI techniques evolved from pure academic research to the edge of commercial applications. New approaches promised to solve some of the inherent problems of traditional software development.

Creating prototypes for various applications became easy using the powerfull harware and software systems supplied by the AI community. The window-oriented, easy to handle user interfaces provided by these systems became a symbol for the superiority of the new approach. New applications were conceivable and the user oriented methods narrowed the gap between programmers and end-users.

The one who did not share the enthusiasm was the system manager. These new systems used strange, expensive hardware or, - if implemented on existing hardware - consumed all the available resources and kept all other users waiting. But implementation on conventional hardware became necessary when AI systems migrated from prototype to operational status.

The integration of new AI solutions into existing conventional systems became a problem.

The following chapters list some of these problems as personal experiences of the author. There may be different ones in other projects, but up to now integration is a problem area.

2 AREAS OF INTEGRATION

Today AI techniques are primarily used in areas where conventional systems failed for various reasons. They are usually part of a more complex system that already exists or is developed in parallel. Parts of the system or interfaces between AI and conventional systems are written in traditional programming languages.

Software engineering therefore contains AI and conventional programming. Integration in this context means integrated software developement and may not be restricted to the technical aspects of various interfaces.

If you look at integration from the software engineering point of view problems may arrise from technical, project management and administrativ issues.

Integrated software development needs solutions for all these issues. But up to now most of the problems are unsolved.

3 TECHNICAL ISSUES OF INTEGRATION

Technical interfaces are a basic requirement for integrated solutions. Combining AI and conventional systems requires interfaces between the hardware and/or software systems.

3.1 Hardware Interfaces

During the late 70th and beginning 80th AI systems were often bound to special AI hardware. LISP or PROLOG were supported as basic languages and the underlying workstation concept asumed one user per system.

Interconnection with conventional hardware was restricted to special hardware and operating systems because appropriate network software was not available. AI systems mostly were restricted to stand alone solutions. Integration had to be done by organisational means. Data transfer was a major problem preventing most prototypes from going into routine use.

Improved network software and the increasing market for unix systems improved the chance of finding a common network protocol for AI and conventional systems. But even in systems developed in the late 80th this problem was solved by introducing a seperate front-end system reponsible for protocol conversion and communication /1/.

New directions look at the AI machine as a special board plugged into a standard bus beside a conventional CPU. Up to now there are no experiences with this approach.

The ideal solution from the customers point of view is integration on top of a common hardware system. This solution requires compatible software structures and efficient implementation of the AI tools and applications.

3.2 Software Interfaces

Most of the AI tools provide some kind of call interface to external languages. It provides access to existing software and data when the AI system is implemented on conventional hardware.

Usually a lot of declarations must be done to set up the interface and the performance at runtime sometimes is pure.

To integrate AI systems with conventional programms three types of interfaces are required:

1. CALL-OUT

   CALL-OUT allows to call programms writen in a conventional language (usually C) from the AI system. It provides access to existing functions and data.

   The CALL-OUT is the common type of interface provided by most of the AI tools.

2. CALL-IN

    CALL-IN allowes to call an AI system from a conventional
    programm as a subfunction. It becomes necessary if the main
    body of a system is a conventional programm but special
    problems are solved by the AI system.

    AI tools usually do not provide this kind of interface.
    Therefore it is difficult to restrict AI systems to tasks they
    are usefull for and to implement the rest in a conventional
    language providing better performance and use of resources.

3. CALL-BACK

    CALL-BACK provides access to local data in the AI system from
    an external function. This means that the external function
    called by a CALL-OUT has free access to data remaining in the
    AI system. If this access is not provided the external
    function is restricted to work on data passed as parameters in
    the call.

    CALL-BACK also should provide functions to create data in the
    AI system. This is usefull to transfer new data from the
    external function to the AI system.

    Access to AI data is usually permitted by a function call
    providing a single piece of data. Compared to data handling in
    conventional languages it is very inefficient.

    When building integrated systems the CALL-BACK funktions are
    used more often than any other type of interface functions.
    The features and the efficiency of that interface are very
    sensitive.

Some standard interface functions should be provided by the AI tool
itself. Database access for instance could be supported on the SQL
level providing direct access to data structures of the AI system. On
top of that, AI systems could provide "database objects" managing
database access as internal functions. These "database objects" also
could solve the conflict that all the AI inference engines expect data
to be completely in their own domain while databases require short and
selective access. Very few AI systems support database access
sufficiently.

If you make use of the interfaces mentioned above you have to develop AI
and conventional software and test it. What you need here is integrated
software development tools and strategies.

But in this point two totally different development strategies meet and
resist integration so far. Interactive, interpretative developement on
one hand and the edit, compile, link and debug cycle on the other.

Usually the external functions as described above are integrated by
linking them to the run-time interpreter of the AI system. The AI
software is loaded on top of the extended run-time interpreter before
each test session. In this process problems arose from the following
deficiencies:

1. Debuggers did not work properly on the large images of the run-time interpreter or were not supported at all. This resulted in the long forgotten test strategy of putting printouts in a programm for test purposes and made testing very inefficient.

2. If the AI system gets larger loading of AI software to the run-time interpreter becomes a time consuming task (more than 3 hours in one case). Usually you can do a lot of testing and development in the AI system after a single load. But if you have errors in an external function you have to reload after each change because your basic run-time interpreter has changed. Testing gets very time consuming then.

3. Writing a testbed for the external functions is expensive because the access to data via CALL-BACK functions must be simulated. Doing this results in a lot of software you have to write for test purposes only, giving you a limited chance of finding errors because many errors are due to data type incompatibility.

There are no integrated development tools so far. Therefore testing integrated systems is one of the most critical and time consuming parts of the project and should be planned carefully.

As mentioned before loading large AI sources files sometimes is a very time consuming task. Not all of the AI tools provide efficient functions to save the image after the sources are loaded or to create incremental run-time versions on conventional hardware.

As a last topic the integration of AI systems with conventional operating systems is stressed.

AI tools are designed to run on special workstations, single user, with as much RAM as possible and a very large disk. Trying to transfer this technology to a conventional, multi user environment causes problems with the amount of resources the system consumes. It seems that normal operating systems are not designed to manage virtual memory spaces of more than 100 MB. But also the CPU load and the amount of disk space is beyond the limitations for conventional systems and frightens any system manager.

Reaching limits conventional systems don't think of AI systems reveal errors in the operating system and the standard tools that have never occured before.

In practice there is no big chance for specialized hardware systems. So AI systems would do better if they did obey the rules of conventional operating systems. May be the next generation systems will.

4 PROJECT MANAGEMENT ASPECT OF INTEGRATION

Building integrated systems means to run projects with conventional and AI parts as mentioned before.

Beside the problem to find an approbriate software engineering tool you have a mixed project team and you have to find a project manager who knows both technologies.

AI developed as a separate research area at universities and the AI community is a closed circle of people. On the other hand only a few engineers from the conventional side know about AI techniques.

Project work at least in the beginning suffers from communication problems and totally different methods of software development.

Usually there is not much experience in integrated software development and it is difficult to find a project manager who knows enough about both technologies to work out an optimal design and test strategy.

It showed usefull to introduce the different technologies in the beginning of the project and to explane and discuss special solutions regularly.

5 ADMINISTRATIV ASPECTS

Integration of AI software into conventional environments is only usefull if the system is used for a longer period.

Beside development the introduction and the usage phase of the system must be supported.

The main recesses of AI in the full range of applications must be seen in the processing of knowledge. In technologically oriented enterprises knowledge is usually subjected to a high dynamic and is influenced by a variety of aspects within and outside the enterprise.

Long term usuage of the AI system usually means permanent update of the knowledge base. AI projects therefore have a different cost structure than conventional projects: the main part of the costs fall to the usage phase of the system caused by permanent actualisations and restructuring of knowledge /2/.

The expense of long term usuage must be carried by the enterprise. This fact has to be seen by the management and return of investment must be calculated in advance.

Therefore the use of AI systems is mostly restricted to strategically important areas of the enterprise.

6 LITERATURE

/1/ James M. Dzierzanowski, Kenneth R. Chrisman, Gary J. MacKinnon, Philip Klahr: The Authorizer's Assistant. A Knowledge-Based Credit Authorization System For American Express. Proc. of the 1989 Conference on Innovative Applications of AI.

/2/ Virginia E. Baker, Dennis E. O'Connor: Expert Systems for Configuration at Digital: XCON and Beyond. CACM Vol 32, No 3 pp 298 - 318.

This series reports new developments in computer science research and teaching – quickly, informally and at a high level. The type of material considered for publication includes preliminary drafts of original papers and monographs, technical reports of high quality and broad interest, advanced level lectures, reports of meetings, provided they are of exceptional interest and focused on a single topic. The timeliness of a manuscript is more important than its form which may be unfinished or tentative. If possible, a subject index should be included. Publication of Lecture Notes is intended as a service to the international computer science community, in that a commercial publisher, Springer-Verlag, can offer a wide distribution of documents which would otherwise have a restricted readership. Once published and copyrighted, they can be documented in the scientific literature.

Manuscripts

Manuscripts should be no less than 100 and preferably no more than 500 pages in length.
They are reproduced by a photographic process and therefore must be typed with extreme care. Symbols not on the typewriter should be inserted by hand in indelible black ink. Corrections to the typescript should be made by pasting in the new text or painting out errors with white correction fluid. Authors receive 75 free copies and are free to use the material in other publications. The typescript is reduced slightly in size during reproduction; best results will not be obtained unless the text on any one page is kept within the overall limit of 18 x 26.5 cm (7 x 10½ inches). On request, the publisher will supply special paper with the typing area outlined.
Manuscripts should be sent to Prof. G. Goos, GMD Forschungsstelle an der Universität Karlsruhe, Haid- und Neu-Str. 7, 7500 Karlsruhe 1, Germany, Prof. J. Hartmanis, Cornell University, Dept. of Computer Science, Ithaca, NY/USA 14853, or directly to Springer-Verlag Heidelberg.

Springer-Verlag, Heidelberger Platz 3, D-1000 Berlin 33
Springer-Verlag, Tiergartenstraße 17, D-6900 Heidelberg 1
Springer-Verlag, 175 Fifth Avenue, New York, NY 10010/USA
Springer-Verlag, 37-3, Hongo 3-chome, Bunkyo-ku, Tokyo 113, Japan

ISBN 3-540-53557-8
ISBN 0-387-53557-8